Environmental Engineering for Civil Engineers

Environmental Engineering for Civil Engineers

Tanjina Nur

ARCLER PRESS

www.arclerpress.com

Environmental Engineering for Civil Engineers

Tanjina Nur

Arcler Press
224 Shoreacres Road
Burlington, ON L7L 2H2
Canada
www.arclerpress.com
Email: orders@arclereducation.com

ISBN: 978-1-77407-623-1 (Hardcover)

ABOUT THE AUTHOR

Tanjina finished her PhD in Civil and Environmental Engineering in 2014 from University of Technology Sydney (UTS). Now she is working as Post-Doctoral Researcher in the Centre for Technology in Water and Wastewater (CTWW) and published about eight International journal papers with 80 citations. Her research interest is wastewater treatment technology using adsorption process.

TABLE OF CONTENTS

LIST OF FIGURES

LIST OF TABLES

LIST OF ABBREVIATIONS

ADB	Asian Development Bank
AfDB	African Development Bank
BACT	best available control technology
BOD	biochemical oxygen demand
BPC	Blended Portland Cement
CFLs	compact fluorescent light bulbs
CO_2	carbon dioxide
COD	chemical oxygen demand
CS_2	carbon disulfide
DAF	dissolved air flotation
DfE	design for the environment
EBRD	European Bank for Reconstruction and Development
EIA	environmental impact assessment
EIB	European Investment Bank
EIS	environmental impact studies
EPA	Environmental Protection Agency
H2O	waters
H_2SO_4	sulfuric acid
HS	hydrogen sulfide
IEE	Initial Environmental Examination
IKEA	Ingvar Kamprad Elmtaryd Agunnaryd
IPS	Intelligent Product System
JBIC	Japanese Bank for International Cooperation
LFT	lake flushing time
MEA	millennium ecosystems assessment
MIT	Massachusetts Institute of Technology
MLSS	mixed liquor suspended solids

NEPA	National Environmental Policy Act
O2	oxygen
O_3	ozone
ODM	Omni diurnal model
PAH	polycyclic aromatic hydrocarbons
PAN	peroxyacetyl nitrate
RDED	Rio Declaration on Environment and Development
SDWA	Safe Drinking Water Act
SMM	sustainable materials management
SPM	suspended particulate matter
TOC	total organic carbon
ToR	terms of reference
UNCED	United Nations Conference on Environment and Development
UV	ultraviolet light
VMT	vehicle mile traveled
VOC	volatile organic compounds
VRR	vegetation recovery rates
WB	World Bank
WHPA	wellhead protection area

PREFACE

Everything appears to matter a lot in the field of environmental engineering. Similar to classical engineering principles, humanities, and social sciences are also equally important to environmental engineering. Most of the environmental engineers see this amalgam of disciplines and skills both rewarding and challenging. In education and research institutes, students and researchers are required to go beyond the boundaries of departments and disciplines. In this way, they can achieve a true combination of skills and disciplines.

The text of this book is structured into arenas essential to all engineering and scientific disciplines. For instance, knowledge about air quality, water resources, hazardous waste, and solid wastes is included in this book. Chapters focusing on risk analysis and impact assessment are also appended in the book. This book essentially contains principles and topics which form the basis of environmental engineering. The emerging environmental problems such as global climate change and green manufacturing technologies are also discussed in this book.

This book essentially provides a fundamental overview of the environmental pollution (e.g., air pollution, water pollution, etc.), and emphasizes on plausible solutions pertaining to environmental engineering. Different chapters of this book provide an understanding of measurement, prevention, and control of different pollution types including air pollution, water pollution, and land pollution. This book also discourses the causes and effects of different environmental problems, and presents solutions to alleviate them and apply procedures to help avoid future damages.

There are eight chapters in this book. The first chapter of the book introduces the readers with fundamental concepts of civil engineering. Chapter 2 focuses on different aspects of environmental engineering. Chapter 3 introduces the readers with different green and environment-friendly technologies which can replace conventional technologies to provide a sustainable environment to the living beings. Chapter 4 focuses on the prevention and management of environmental pollution while Chapter 5 discusses the principles of air quality control.

Chapter 6 briefly discusses the fundamentals of water pollution and its control. Chapter 7 illustrates the applications of environmental engineering in different civil architectures for creating a sustainable environment. Finally, Chapter 8 describes the fundamentals of environmental impact analysis.

The contents of this book are mainly intended for a diverse range of audiences including engineers, scientists, environmentalists, policymakers, and students. However, this book can be regarded as a reference guide for students and researchers of environmental engineering and civil engineering.

Author

CHAPTER 1

Fundamentals of Environmental Engineering

CONTENTS

1.1. INTRODUCTION

The profession of Civil Engineering is related to the constructed environment. Civil engineers design, plan and construct main facilities, which include, transit systems, dams, highways, airports, energy facilities, buildings, water and wastewater treatment systems, canals, bridges and harbors. Civil engineers mainly take responsibility to manage our water, air and energy resources and also protect our society from natural disasters such as earthquakes, floods, etc. and also protect from the threats produced by the society such as toxic wastes (Chadwick & Morfett, 2002).

Because these tasks are critical to the day-to-day lives of a lot of people and the services involved are actually significant, civil engineers have a significant responsibility to the public. The role of civil engineers is not only technical but it is more than that which requires a high degree of communication skills and capability to deal with the people. Civil engineers have their jobs in consulting firms, the industry as well as in the government sector. Civil engineering is one of the few areas of engineering in which dealing with the public and public agencies are directly involved in every stage of main infrastructural projects (Morse et al., 1972; An et al., 2008).

For the protection and control of the environment and its resources, sustainable designs are involved in Environmental Engineering. Systems are designed by environmental engineers for water treatment and quality, hazardous waste management, wastewater treatment and control of air pollutants. Although the phrase "environmental engineering moderately new, Environmental Engineering has a long history. Before delving into the particular examples and the nitty-gritty of ideas and designs, it is useful to review briefly that history and stare at what the upcoming holds (Perry & Perry, 1976; Abraham, 2005).

The origins of environmental engineering range back to the start of civilization. Managing waste and providing clean water became essential whenever people assembled in ordered settlements. For antique cities, the accessibility of a reliable water source often intended the difference between existence and obliteration, and water supply became a self-protective requirement. The same people who were called on to build the moats and walls were the builders of the aqueducts and wells as well as the engines of the war and catapults (Ottosen et al., 2008). These men were the engineers of ancient times. After the mid-1700s the engineers primarily engaged in matters of warfare began to distinguish themselves from the engineers who built facilities for the civilians and these engineers were categorized in the term "Civil Engineering." In the early years of the United States, engineers were mostly trained at the newly formed United States Military Academy or self-educated. Civil engineers who were the builders of the buildings, roads, railroads, and bridges were called on the construct and design water supplies for the cities, and to deliver suitable systems for the management of storm water and waterborne waste (Bielefeldt, 2011, 2013).

Because of the lack of waste and water management, the advent of the industrialization carried with it extremely unhygienic situations in the cities. There was no public uproar, though, until it became obvious that water could transmit disease. From that time onward, civil engineers had to ensure that water would not be a source for the disease transportation along with the provision of a suitable supply of water. For civil engineers, public health became the essential concern and commended with delivering water supplies to the population middles, and the removal of waterborne disease became the main aim in the late 19th century. The civil engineers assigned with the provision of clean water supplies and drainage of cities became sanitary engineers (in the United States) and public health engineers (in Britain).

1.2. ENVIRONMENTAL ENGINEERING TODAY

A remarkable decrease in the transmission of acute diseases by the contaminated water and air have been achieved by the sanitary engineers. The severe impacts of pollution are for all intentions and purposes eradicated in the United States (Junwe et al., 2012). These severe concerns have been changed, though, by more chronic and complex problems such as depleting aquifers; global transport of persistent; climate change; bio-accumulating and toxic chemicals; indoor air pollution; synergistic influences of complex mixtures of human-made from domestic products and pharmaceuticals in wastewater effluents; endocrine-disrupting chemicals; rivers and streams; and a lack of info on the consequence on environmental and human health and safety of rapidly developing new materials, such as nanoparticles (Field, 1973; Deng et al., 2003).

Challenges to discrete environmental media such as water and air can no longer be measured and achieved within individual sections. They must be achieved at the ecosystem level to evade flowing pollution apprehensions from one environmental medium to another. To address these long-lasting difficulties before they become severe, engineers and scientists are looking for to comprehend cities, environment, and industry as interrelating systems (i.e., as social systems, interconnected ecosystems, and industrial systems) and to think preemptively and proactively so that we can evade unintentional significances rather than having to achieve them reactively (Gerogiorgis & Jolliffe, 2015).

In most advanced countries nowadays, the public view has changed to where the immediate and direct health influences of environmental pollution are no longer the only concern. The hygiene of watercourses, for the aid of the stream itself, has become a pouring force, and the legislature has been approved addressing our wish for a hygienic atmosphere (Cooke & Vermaire, 2015). The health of ecosystems, the conservation of species and the defense of wildlife territory, have become effective purposes for the expenditure of capitals. Such wisdom of mission, frequently mentioned as an *environmental ethic*, is a main dynamic force behind contemporary environmental engineering and is required by the public as a public worth (Forthman & Ogden, 1992).

Figure 1.1: *Excellent display of civil-environmental engineering collaboration.*

[Source: http://greenchemistry.alliedacademies.com/2019/events-list/civil-environmental-engineering]

In the 20th century, an environmental code was frequently potholed against the wishes of those who desired to exploit natural assets for human gain. Common thinking expected that a trade-off had to be made: One had to select between the environment *or* the economy (Jones et al., 2002; Xiao-Zhou, 2007).

1.2.1. Environmental Engineering on the Horizon

In the 21st century, it is obvious that the natural capital and ecosystems on our planet are not unlimited. Maintaining and preserving the health, social well- being and economic of people rest on maintaining and preserving the reliability of the ecosystem services they provide and ecosystems itself. The answer is not a trade-off (Grose, 2006). The answer is the well-being of the environment and the economy. A common aim has appeared throughout the world—the aim of maintainable growth—explained by a 1987 United Nations commission in the Brundtland Report as "growth that meets the requirements of the present without negotiating the aptitude of future peers to meet their own desires." Maintainable development (sometimes mentioned as "sustainability") means dissimilar things to dissimilar people and communities liable on the nature of their actions and the geographic, cultural, environmental, and economic settings in which they function (Socolofsky et al., 2011; Thibodeaux et al., 2011).

Environmental engineering was a responsive profession in the past, reacting to the complications produced by the rise in our standard of living and the development of world populations. The future complications that environmental engineers will discourse can be inferred. For example, we know, the contamination problems were local problems only a few decades ago and now are worldwide in scale. We also know that the discharge and continued use of old and new chemicals into the environment will have random and sometimes synergistic impacts (North et al., 2011; Leal Filho, 2012). And we know that concentrating on energy efficiency and other efficiency developments alone will not sufficiently address developing reserve restraints as populations cultivate and the quality of life in emerging countries advances. Environmental engineers, now and in the future, will play a key role in understanding maintainable growth (Luthy & Small, 1990).

Unnecessary to say, traditional hygienic engineering education founded on public health, practical hydraulics, chemical engineering processes is insufficient to deal with these complicated matters. The new type of environmental engineers commended with the defense of the environment human health will hold the natural sciences and will pact on a cutting-edge level with the application of chemical and biological sciences, including biotechnology, nanotechnology, chemical fate, information technology, toxicity, and life cycle influence deliberations (Abbott, 2001). Environmental engineers will also advance greater sympathetic of industrial procedures, including product development and design. And ultimately, because these complications are all in the public field, they will acquire to put on the social sciences such as public policy, economics, and communications and will acquire to work with varied stakeholders to solve complications. Since all of these matters are improbable to be completely addressed in present civil or environmental engineering prospectuses, students may want to pursue contact with them in sister academic programs, such as environmental science and management, business management and public policy (Masuda, 1988).

Environmental engineers will practice their design exercise to be preemptive and proactive in the growth of solutions. As they move far from a focus on end-of-pipe action and even outside pollution inhibition through manufacturing controls, they will move to the use of a project to avoid complications from the start. Emerging fields such as green chemistry, design and green engineering for the environment, discoursed at numerous points in this textbook, will help conservational and other engineers grow maintainable products and substitute sustainable materials managing (Jingjiao, 1988).

As the complications handled by environmental engineers endure to produce more complex, there is a raising need for frameworks and principles to the director the growth of solutions. In 2003, around 65 scientists and engineers assembled in Sandestin, Florida to grow a set of ideologies for green engineering. The U.S. Environmental Protection Agency (EPA) explains green engineering as "the commercialization, design, and use of products and processes [that] are economical and feasible while minimalizing: (1) production of contamination at the source, and (2) danger to the environment and the human health." The group perceived a set of values that went outside what was characteristically understood as the scope of green engineering to discourse social elements. As such, they became recognized as the Sandestin Principles for Sustainable Engineering (Sandestin Principles). Other groups of principles have been established to upkeep the development and design of processes and products with advantages for the environment and human health. These comprise the 12 Principles of Green Engineering and the 12 Principles of Green Chemistry. The USEPA has approved the nine Sandestin Principles as the Principles of Green Engineering. Composed, these principles deliver a broad framework and a guide for all engineers who pursue to aid solve the complications of the 21st century. They describe well the increasing role of the environmental engineer.

Sustainable engineering alters current engineering practices and disciplines into those that encourage sustainability. Sustainable engineering includes implementation and the development of economically and technologically feasible products, systems and processes that encourage human welfare while shielding human health and uplifting the defense of the biosphere as a standard in engineering resolutions (Abraham, 2006a; 2006b; García-Serna et al., 2007).

These principles can assist as a director for environmental engineers in all kinds of employment whether in consulting, government, private industry or academia. The public appreciates and distinguishes the work of the environmental engineer and is ready to use societal capitals to attain sustainability. The specialized environmental engineering community must prepare now to meet this trial, confirming that environmental engineers endure attaining the depth of technical know-how characteristically expected of environmental engineers and completed with the aptitude to understand complications at the system level and to cooperate effectively with experts and non-experts from other sectors and disciplines.

1.2.2. Sandestin Principles for Sustainable Engineering

Sandestin principles are listed below (Abraham & Nguyen, 2003; Anderson et al., 2008):

i. Engineer products and processes holistically, integrate environmental impact assessment tools using systems analysis.
ii. Improve and conserve natural ecosystems while shielding human well-being and health.
iii. In all engineering activities, use life-cycle thinking.
iv. Ensure that all energy inputs and outputs and materials and areas integrally safe and benign as possible.
v. Minimize the reduction of natural resources.
vi. Struggle to avoid waste.
vii. Design, develop and apply engineering solutions while being cognizant of local geography, cultures, and aspirations.
viii. Generate engineering solutions outside dominant or current technologies; innovate, improve and create (technologies) to attain sustainability.

ix. Actively involve stakeholders and communities in the growth of engineering solutions.

1.3. EXAMPLES OF ENVIRONMENTAL PROBLEMS

The following are samples of environmental complications that have been recognized and solved. They exemplify some of the controversies and principles intrinsic in the area of environmental engineering.

1.3.1. The Holy Cross College Hepatitis Outbreak

Members of the 1969 Holy Cross College football team got sick, following the Dartmouth game. They had abdominal pain, nausea, and high fever and were becoming cynical all features of catching hepatitis. Through the next few days over 87 affiliates of the football program players, trainers, coaches, and other personnel became ill. The college annulled the rest of the football season and became the emphasis of an epidemiological secret. How could a whole football team have contracted to catch hepatitis?

The disease is supposed to be conveyed typically from person to person, but rashes can also occur, often due to polluted water supplies or seafood. There are many kinds of hepatitis viruses, with extensively fluctuating impacts on humans (Wacker et al., 1972; Friedman et al., 1985). The slightest lethal is Hepatitis A virus, which consequences in numerous weeks of feeling unwell and rarely has lasting impacts; Hepatitis B and C, though, can consequence in severe problems, particularly liver damage, and can last for many years. At the time of the Holy Cross epidemic, the hepatitis virus had not been isolated and little was known of its etiology or effects (Bryan et al., 1974).

When the college became conscious of the significance of the epidemic, it requested and established help from federal and state agencies, which sent epidemiologists to Worcester. The epidemiologists' first task was to accumulate as much info as possible about the affiliates of the football team— where they had gone, who they had been with, what they had drunk and what they had eaten. The aim was then to assume from the signs how the epidemic had happened. Some of the info they knew or found out was as below (Taylor, 1972):

i. Since the development phase of hepatitis is about 25 days, the contamination had to have happened erstwhile before 29 August or around there.

ii. Football members who left the team before 29 August were not diseased.

iii. Varsity players who reached late, after 29 August, likewise were not infested.

iv. Freshman football players reached on 3 September, and none of them got the infection.

v. Both the freshmen and varsity players used the same dining facilities, and since none of the freshmen became infected, it was unlikely that the dining facilities were to blame.

vi. A coach who established hepatitis did not eat in the eating room.

vii. There was no communal thread of the players having plagued at restaurants where polluted shellfish might have been the cause of the virus.

viii. The kitchen arranged a mixture of sugar, ice, honey and water for the team but since the kitchen workforce tested this drink before and after going to the exercise field and consequently none of the kitchen workforce established hepatitis, it could not have been the drink.

The lack of replacements enforced the epidemiologists to emphasize on the water source. A buried line delivers water from Dutton Street which is a dead-end street, to the football training field, and the college obtains its water from the city of Worcester, where a drinking fountain is used during training. Samples of water taken from that fountain displayed no adulteration. The nonexistence of pollution during the selection did not, though, rule out the likelihood of infection spread through this water line. The line ran to the training field through a meter pit and a sequence of recessed sprinkler boxes used for sprinkling the field (Figure 1.2).

Two other bits of info came out to be critical. One of the houses on Dutton Street was observed to have children who had hepatitis. During the summer evenings, the kids played near the sprinkler boxes and frequently unlocked those, wallowing round in the small ponds shaped in the pits (Craun et al., 1976). But if the children had polluted it, how did the water in the play ponds get into the waterline with the line continually under positive pressure?

The ultimate piece of the enigma fell into place when the epidemiologists observed that during the evening of 28 August a great fire had happened in Worcester enduring well into the initial hours of the next day. The request for water for this fire was so excessive that the habitations on Dutton Street observed themselves deprived of any water pressure at all (Meyers et al., 1974). That is, the pumpers tapping out the fire were propelling at such a high rate that the pressure in the water line converted negative. If then, the kids had left some of the regulators in the sprayer boxes exposed and if they had polluted the water around the box, the hepatitis virus must have arrived the drinking water line. The next morning, as pressure was recommenced in the water lines, the polluted water was strapped to the distant end of the line—the drinking cascade on the football ground—and all those coaches players, and others who drank from the drinking fountain were infected with hepatitis (Chang & O'Brien, 1970; Crusberg et al., 1978).

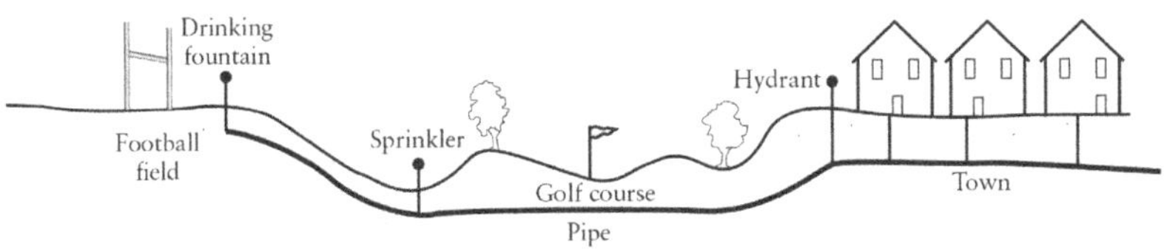

Figure 1.2: The pipeline that carried water to the Holy Cross practice field came from Dutton Street and went through several sprinkler boxes.

[Source: https://www.amazon.com/Introduction-Environmental-Engineering-Aarne-Vesilind/dp/0495295833]

This situation establishes a typical *cross-connection*, or the physical contact between contaminated water and treated drinking water and the possibly severe significances of such influences. One of the aims of environmental engineering is to plan systems that defend public health. In the circumstance of piping, engineers need to plan systems in such a method as to evade even the likelihood of cross-connections being produced, though as the Holy Cross College event displays, it is improbable that all potentials can be expected (Hoffman et al., 1979).

1.3.2. The Disposal of Wastewater Sludge

The well-known American writer and linguist H. L. Mencken (1945), in his article *The American Language: An Inquiry into the Development of English in the United States* (Alfred A. Knopf Inc. 1977), detected that various fresher words in our language have been made as a grouping of sounds that in themselves transport a meaning or a picture. For instance, "crud" initiated out as C.R.U.D., lingering recurrent unspecified dermatitis, a medicinal diagnosis for American soldiers posted in the Philippines in the initial 1900s. The word has a picture lacking a definition. Try laughing, and in your sugary, pleasantest way, say "crud." It just can't be completed. It always sounds... well ... cruddy.

A grouping of consonants that Mencken points out as being mostly unpleasant is the "sl" sound. Skimming the vocabulary for words preliminary with "sl" yields slimy, slug, sludge slut, slum, slovenly and, of course, slither. The very sound of the word is unpleasant, so the stuff must be somewhat else!

Sludge is created in a wastewater management plant as the deposit of wastewater management. Wastewater management plants waste energy as humans are not efficient users of the chemical energy they consume. And similar to the human body, the digestion of the wastewater management plant is not efficient (Davis & Hall, 1997; Ødegaard et al., 2002). While these plants yield pure water that is then predisposed into the adjacent waterway, the plants also foodstuffs an offshoot that still has considerable chemical energy. This remainder cannot be merely predisposed because it would simply overcome a marine ecology or cause irritation difficulties or even be dangerous to human fitness. The disposal and treatment of wastewater sludge are one of the environmental engineering's greatest pressing glitches. (To decrease the negative public view of sludge removal, one water quality connotation recommended that the stuff exiting the management plant be called "bio solids" as an alternative of "sludge." Of course, "a rose by any other name").

Because sludge removal is so problematic and because inappropriate removal can origin human health issues, governmental guidelines are wanted. In the situation of environmental guidelines by governmental assistance, human well-being or health is frequently considered against economic deliberations (Kelessidis & Stasinakis, 2012). That is, how much are we ready to pay to have an improved environment? The supposition, or at slightest faith, is that the regulating support has the info essential to regulate just what consequence certain regulations will have on human health. Inappropriately, this is rarely the case, and governing agencies are enforced to make choices on the basis of scarce or unavailable scientific info. The controller must equilibrium diverse constituents and competing interests. (For instance, in Iceland the existence of elves has been taken consciously, and roads have been redirected to stop injury to the alleged homes of the little people!) In a democracy, the controller signifies the benefits of the public. If the rules are considered intolerable, the public can change the rules (or can change the controller!).

A case of a disliked rule in the United States was the 55 mph speed limit on regional roads, a rule that was normally overlooked and ultimately canceled. The U.S. Department of Transportation regulation creators miscalculated the readiness of the public to slow down on highways. The two advantages—saving lives and a decrease in gasoline use were estimable goals, but the rule was banned because it required too much of the public. In the situation of the speed limit, transport engineers were capable to state clearly that the dropping of the speed limit from 65 mph to 55 mph would save about 20,000 lives yearly, but this advantage did not influence public view. The public was not ready to pay the value of lesser highway speeds (Epstein et al., 1976; Hossain et al., 2009).

Environmental regulations likewise seek commendable and ethically justifiable aims, typically the improvement of public health (or dealing with the negative, the deterrence of disease or early death). Environmental regulations need the regulator to weigh the advantages accumulated by the rules against the costs of the rules. Frequently the worth of human health guard is stable against the burden of regulatory movements that may involve economic costs and limits on liberty by restraining contaminating behaviors (Tay & Show, 1997). That is, the controller, by setting environmental rules that improve the health of the public, takes away liberty from those who would release contaminants into the environment. The controller stabilizes the good of public health in contrast to the cost of wealth or freedom—in consequence dropping freedom and taking affluence. Setting unembellished limits on releases from civic wastewater treatment plants needs that public taxes be elevated to pay for the extra management. Elimination of the release of a heavy-metal-laden manufacturing sludge needs establishments to set up luxurious pollution-prevention systems and avoids them from clearing these wastes by least-cost means. Setting severe intake water values likewise consequences in better disbursement of throwaway wealth in structure better water management plants. In every situation, the controller, when setting environmental rules, balances the moral value of taking wealth against the moral value of public health. Thou shalt not hurt versus thou shalt not steal. This is a moral predicament, and this is precisely what the controller faces in setting environmental rules (Liu, 2003).

Earle Phelps was the first to identify that maximum environmental controlling choices are made by means of what he named the *principle of expediency* (Phelps, 1948). A sanitary engineer recognized for his work with the development of the Streeter and watercourse sanitation–Phelps dissolved oxygen sag curve equation (Chapter 8), Phelps labeled expediency as "the effort to decrease the arithmetical measure of likely damage, or the rational measure of present hazard, to the bottom level that is feasible and practicable within the limits of financial incomes and engineering ability." He accepted that "the ideal or optimal ailment is seldom available in practice, and that it is wasteful and consequently inexpedient to need a nearer method to it than is willingly accessible under present engineering practices and at justifiable budgets." Most prominently for today's standard setters, who frequently have a problem protecting their choices, he guided that "the code of expediency is the rational foundation for managerial standards and should be honestly stated in their protection." Phelps saw nothing incorrect with the usage of standards as a type of speed limit on contamination disturbing human health. He also unstated the laws of lessening returns and lag time for practical possibility. Yet he always strapped toward plummeting environmental hazards to the lowermost practical levels.

The obligation of the controller is to include the finest accessible science into governing judgment making. But issues arise when only inadequate scientific info is accessible. The unpredictability of the ecological impact of sludge on human well-being prompts scientific vulnerability and makes sludge disposal problematic. The issue in creating sludge transfer guidelines is that wastewater sludge has obscure and dynamic properties and carries on distinctively in various environmental media. Controllers must decide when the presence of sludge is hazardous and what should and should be possible about it.

Even with such complexity, in the mid-1980s the USEPA started a program to create health-based sludge disposal guidelines. The organization held up as long as it could, despite the fact that they were commanded by the 1972 Clean Water Act to set such guidelines. The assignment was overwhelming, and they knew it. They set about it in an intelligent manner, first determining every one of the methods by which the constituents of sludge could harm people and afterward defining the most pessimistic scenario situations. For instance, for sludge burning they accepted that an individual lives for a long time promptly downwind of a sludge incinerator and inhales the discharges 24 hours of the day. The individual never moves, the breeze never moves, and the incinerator continues transmitting the contaminants for a long time. Of most alarm would be volatile metals, for example, mercury. Utilizing epidemiological proof, for example, from the

Minamata catastrophe in Japan, and extrapolating a few sets of extent, the USEPA assessed the all-out passable discharges of mercury from a sludge incinerator.

By building such most worst-case scenario yet absolutely ridiculous situations, the USEPA built up a progression of draft sludge disposal guidelines and distributed them for open remark. The reaction was quick and overpowering. They got more than 600 official reactions, practically every one of them criticizing the assumptions, the procedure, and the conclusions. A significant number of the discourses brought up that there are by and by no known epidemiological information to show that appropriate sludge disposal is in any capacity harmful to general society. Without such data, the setting of exacting gauges appeared to be ridiculous (Singh et al., 2004).

Pounded by such unfavorable responses, the USEPA deserted the health-based methodology and embraced Phelps-type expediency guidelines that define two sorts of sludge, one (Class B) that has been treated by such methods as anaerobic digestion and the other (Class A) that has been sanitized. Class A sludge can be discarded on all farmland, yet Class B sludge has limitations, for example, holding up 30 days before dairy cattle could be reintroduced to a field on which sludge had been splashed. The sludge that has not been dealt with (probably Class C, despite the fact that this is not so assigned) is not to be discarded into the environment. This guideline is practical on the grounds that all wastewater treatment plants in the United States currently have some sort of sludge adjustment, for example, anaerobic digestion, and a guideline that the vast majority of the treatment plants are now consenting to is a popular guideline.

The lack of helpful epidemiological data on the impact of sludge constituents on human well-being constrained the USEPA, in building up their worst-case scenario, to blunder such a great amount on the conservative side that the guidelines got unrealistic and would not have been acknowledged by general society. The ruin of the health-based guidelines was that the controllers could not state what number of individuals would be harmed by sludge disposal that did not meet the projected criteria. Without such data, general society concluded that it essentially would not like to be burdened with what they saw as pointless guidelines. The USEPA would have been taking a lot from them (cash) and giving back an undefined and obviously minor benefit (health). So the USEPA chose to do what was convenient—to have the wastewater management plants do what they can, (e.g., anaerobic digestion in some situations or cleansing by heat in others), realizing that these guidelines would at present be better to none at all. As our expertise at treatment improves and as we choose to spend more cash on wastewater treatment, the norms can be fixed in light of the fact that this would then be morally convenient.

Administrative decision making, for example, setting sludge disposal guidelines, has moral ramifications since it includes disseminating benefits and costs among influenced residents. The guideline of expediency is a model that requires a controller to enhance the benefits of health security while limiting expenses inside the limitations of technical feasibility. Phelps' expediency guideline, proposed more than 50 years prior, is as yet a helpful utilization of morals utilizing scientific information to set dynamic but then enforceable ecological guidelines. In the situation of sludge disposal, the USEPA settled on an ethical choice dependent on the rule of expediency, weighing the moral good of human health security against the ethical harm of taking the wealth by requiring expensive wastewater muck treatment and disposal.

1.3.3. The Donora Episode

It was a typical Western Pennsylvania fall day in 1948, shady and still (Shrenk et al., 1949). The occupants of Donora, a little plant town on the banks of the Monongahela River, did not give a lot of consideration to what seemed, by all accounts, to be a particularly smoggy climate (Ciocco & Thompson, 1961). They had seen more terrible. Some even recalled days when the air was dense to such an extent that streamers of carbon would really be visible, lingering palpably like dark icicles. So the youngsters' Halloween march went on as booked, as did the secondary school football match-up Saturday evening, in spite of the fact that the mentor of the rival group pledged to fight the game. He guaranteed that the Donora mentor had thought up to have a pall of brown smog remain over the field so that, if a forward pass were tossed, the ball would totally vanish from seeing and the receivers would not know where it would return (Amdur & Corn, 1963; Snyder, 1994).

In any case, this was not the same as the typical smoggy day. By that night 11 individuals were dead, and ten more were to bite the dust in the following few hours. The smog was dense to the point that the specialists treating patients would get lost moving from house to house. By Monday practically a large portion of the individuals in the community of 14,000 was either in emergency clinics or ill in their own homes with serious vomiting, headaches, and cramps. Pets endured most, with every one of the canaries and a large portion of the cats and dogs dead or biting the dust. Indeed, even houseplants were not invulnerable with the impacts of the brown smog (Townsend, 1950).

There were insufficient emergency vehicles or medical clinics ready to aid a calamity of this extent, and numerous individuals passed on for the absence of prompt care. Firefighters were conveyed with tanks of oxygen to do what they could to help the most gravely sick. They needed more oxygen for everybody, so they gave individuals a couple of breaths of oxygen and proceeded to help others (Korshover & Angell, 1982).

At the point when the environment finally cleared on 31 October, six days of serious poisonous smog had incurred significant damage, and the full extent of the scene (as these air quality catastrophes came to be known) got apparent. The exposure encompassing Donora introduced another mindfulness and pledge to control air quality in our networks. Health laborers conjectured that, if the smog had proceeded for one more night, very nearly 10,000 individuals may have passed on.

What is so uncommon about Donora that made this episode conceivable? To start with, Donora was a traditional steel belt plant town. Three huge modern plants were on the waterway—a steel plant, a wire factory, and a zinc plant for galvanizing the wire—the three together creating galvanized wire. The Monongahela River gave transportation to world markets, and the accessibility of crude materials and trustworthy work (frequently imported from Eastern Europe) made this a most profitable endeavor. During the end of the week when the air quality circumstance around got critical, the plants did not hinder production. Clearly, the plant chiefs did not detect that they were in any capacity liable for the state of the residents of Donora. Just Sunday night, when the full degree of the disaster got known, did they shut down the heaters?

Second, Donora sits on a twist in the Monongahela River, with high precipices to the outside of the twist, making a bowl with Donora in the center (Figure 1.3). On the night of 25 October 1948, a reversal condition subsided into the valley. This meteorological condition, making them nothing to do with contamination, just constrained the upward development of air and made a kind of lid on the valley. Contaminations transmitted from the steel plants, therefore, could not get away and were caught under this lid, delivering a relentlessly expanding degree of contaminant focuses (Winkelstein, 1972).

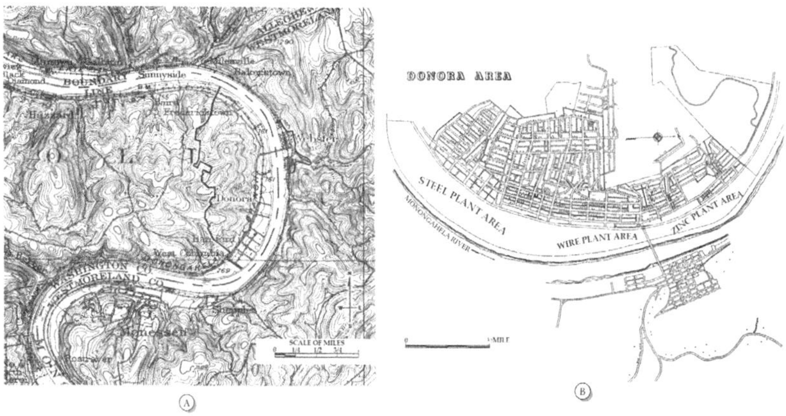

Figure 1.3: Donora was a typical steel town along the Monongahela River, south of Pittsburgh, with (A) high cliffs creating a bowl and (B) three steel mills producing the pollutants.

[Source: https://www.amazon.com/Introduction-Environmental-Engineering-Aarne-Vesilind/dp/0495295833]

The steel organizations demanded that they were not to blame, and in reality, there never was any deficiency suggested by the uncommon investigation into the incident. The organizations were working inside the law and were not pressuring any of the laborers to work in their plants or anybody to live in Donora. Without legislation, the organizations felt no commitment to pay for air contamination gear or to change procedures to decrease air contamination. They accepted that, if just their organizations were required to pay for and introduce air contamination control hardware, they would be at a focused detriment and would inevitably leave the business.

The catastrophe constrained the State of Pennsylvania and in the long run the U.S. government to act and was the single most prominent catalyst to the section of the Clean Air Act of 1955, in spite of the fact that it was not until 1972 that viable federal legislation was passed. In Donora and close by Pittsburgh, in any case, there was a feeling of forswearing. Smoke and poor air quality comprised a sort of macho condition that implied occupations and flourishing. The Pittsburgh press gave the updates on the Donora catastrophe equivalent charging to a jail breakout. Indeed, even in the mid-1950s, there was a dread that, if individuals fought about contamination, the plants would shut down and the employments would vanish. And surely, the zinc plant (thought to be the fundamental culprit in the arrangement of the poisonous smog) shut down in 1957, and the other two factories shut 10 years after the fact. Donora, be that as it may, lives on as the area of the absolute most significant scene that put into a movement our present pledge to clean air.

1.3.4. Jersey City Chromium

Jersey City, in Hudson County, New Jersey, was at one time the chromium handling capital of America, and throughout the years, 20 million tons of chromite mineral preparing buildup were sold or parted with as fill (Bartlett & Vesilind, 1998). There were many defiled locales, including ball fields and storm cellars hidden in the two homes and businesses. It was normal for splendidly hued chromium mixes to take shape on clammy storm cellar dividers and to blossom on soil surfaces where soil dampness dissipates, making something like orange hoar ice of hexavalent chromium—Cr (VI). A messed up central conduits in the wintertime brought about the arrangement of brilliant green ice because of the nearness of trivalent chromium—Cr (III).

The organizations that made the chromium squander issue never exist, yet three aggregates acquired the obligation through a progression of takeovers. In 1991, Florence Trum, a neighborhood occupant, effectively sued Maxus Energy, a backup of one of the aggregates, for the passing of her significant other, who stacked trucks in a distribution center fabricated legitimately over a chromium waste disposal site. He built up an opening at the top of his mouth and cancer growth of the thorax; it was controlled via postmortem examination that chromium harming caused his death. While the subsidiary organization did not create the chromium pollution, the judge decided that they thought about the risks of chromium (Burke et al., 1991; Freema et al., 1997).

The State of New Jersey at first burned through $30 million to find, unearth, and expel a portion of the tainted soil. In any case, the degree of the issue was overpowering, so they halted these endeavors. The chief of harmful waste cleanup for New Jersey conceded that, regardless of whether the dangers of living or working close to chromium were known, the state did not have the cash to expel it. Starting appraisals for site remediation were well over $1 billion.

The residents of Hudson County were angry and apprehensive. Those wiped out with malignant growth thought about whether it could have been forestalled. Mrs. Trum saw the culprits to dressed specialists who were eager to take risks with other people groups' lives. "Big business can do this to the little man," she said.

The pollution in Jersey City was from ventures that pre-owned chromium in their procedures, including metal plating, textile manufacturing, and leather tanning. The accumulation of this chrome in dumps brought about chromium-sullied water, sludge, and soils. Chromium is especially difficult to control due to the multifaceted nature of its chemical behavior and lethality, which converts into scientific vulnerability. Vulnerability compounds the propensity of regulatory agencies to make traditionalist and defensive presumptions, the inclination of the managed to scrutinize the scientific reason for guidelines, and the inclination of conceivably presented residents to fear potential hazards (Lioy et al., 1992; Freeman et al., 1995; 2000).

Chromium exists in nature principally in one of two oxidation states—Cr (IY) and Cr (III). In the reduced type of chromium, Cr (III), there is a propensity to frame hydroxides that are moderately insoluble in water at neutral pH esteems. Cr (III) does not seem, by all accounts, to be cancer-causing in creatures and bioassays. Naturally, complexes Cr (III) got one of the more well-known dietary enhancements in the United States and can be acquired monetarily as chromium picolinate or with exchange names like Chromalene to help with legitimate glucose digestion, weight reduction, and muscle tone (Tai & Jou, 1999).

At the point when oxidized as Cr (VI), nonetheless, chromium is profoundly lethal. It is ensnared in the advancement of lung cancer and skin sores in industrial laborers. As opposed to Cr (III), all Cr (VI) mixes have been demonstrated to be powerful mutagens. The USEPA has classified chromium as a human cancer-causing agent by inward breath dependent on proof that Cr (VI) causes lung cancer. In any case, chromium has not been demonstrated to be cancer-causing by ingestion (Witmer et al., 1991).

What entangles chromium science is that, under certain ecological conditions, Cr (VI) and Cr (III) can interconvert. In soils comprising of manganese, Cr (III) can be oxidized to Cr (VI). While the organic matter may difficult to diminish Cr (VI), it might likewise complex Cr (III) and make it increasingly dissoluble, encouraging its movement in groundwater and increasing the probability of experiencing oxidized manganese present in the dirt. Given the heterogeneous idea of soils, these redox responses can happen at the same time (Stern et al., 1992). Cleanup limits for chromium were initially founded on contact dermatitis, which was disputable. While some see contact dermatitis as a real case to harm, others facetiously proposed administrative limits for poison ivy, which likewise causes contact dermatitis. The philosophy by which dermatitis-based soil limits were resolved went under assault by the individuals who scrutinized the legitimacy of skin fix tests and the inductions by which fix test results convert into soil Cr(VI) levels.

Through the discussion, there developed some helpful innovations to help in the resolution of the debates. For instance, diagnostic tests to measure and differentiate Cr (IV) and Cr (III) in soils were created. Prior to the entire existence of New Jersey's chromium issue, these examine were untrustworthy and would have required re-mediating soil dependent on complete chromium. Other specialized/scientific progresses included remediation techniques intended to artificially diminish Cr (VI) to Cr (III) to lessen chance without unearthing and expulsion of soil assigned as unsafe waste (Kitsa et al., 1992; Wedeen et al., 1996).

The dissatisfaction with moderate cleanup and what the residents saw as a twofold talk by the researcher's finally finished in the abnormal step of correcting the state constitution to give assets to hazardous waste cleanups.

Figure 1.4: A scene from Erin Brockovich, starring Julia Roberts and Albert Finney.

[Source: https://www.amazon.com/Introduction-Environmental-Engineering-Aarne-Vesilind/dp/0495295833]

Chromium is likewise the culprit in the profoundly fruitful film Erin Brockovich, featuring Julia Roberts and Albert Finney (Figure 1.4). Erin Brockovich (Julia Roberts) was a committed and eager open supporter, unsophisticated in lawful amenities, who helped win a significant settlement in the contamination of groundwater with chromium around a mechanical site.

1.3.5. The Discovery of Biological Wastewater Treatment

Prior to 1890, substance precipitation with land cultivating was the standard technique for wastewater treatment in England (HG, 2001). The most well-known alternative was to first enable the waste to go anaerobic in what we today call septic tanks. Such putrefaction was believed to be a simple chemical process in light of the fact that the physical nature of the waste clearly changed (Lu et al., 2014). The effluent from the septic tanks was then synthetically precipitated, and the sludge was applied to farmland or shipped by exceptional sludge boats to the sea. The somewhat treated effluent was released to streams, where it, for the most part, made extreme odor issues (Narayanasamy et al., 2015).

At the time London's administrations were given by the London Metropolitan Board of Works, which, among its different duties, was charged with tidying up the River Thames. The main architect for this association was Joseph Bazalgette, who moved toward the Thames water quality issue based on what was at the time a splendidly discerning designing point of view. In the event that the issue was an awful smell in London, why not fabricate long interceptor sewers along the two banks of the Thames and release the wastewater far downstream (Figure 1.5)? Although costly, this arrangement was received and the city went through enormous entireties of cash to send out the wastewater to Barking Creek on the north bank and Crossness Point on the south bank. The thought was to gather the sewage at these focal areas and afterward treat it to deliver a valuable item, for example, fertilizer. None of the reusing plans worked out as expected and the wastewater was released untreated from the outfalls into the lower Thames. Since at the area of the outfalls the Thames is a tidal estuary, the underlying arrangement was to release wastewater just during the active tide (LaPara et al., 2002; Daims et al., 2006). Lamentably, the waste-water must be released continuously, and the approaching tide conveyed the malevolent smelling stuff back up to the city and put great pressure on the legislators to accomplish something.

Figure 1.5: Early sewer construction.

[Source: http://www.unmuseum.org/7wonders/london_sewers.htm]

A few arrangements were considered. One was just to proceed with the interceptor sewers and extend them right toward the North Sea, yet this ends up being restrictively expensive. Another arrangement was to shower the wastewater ashore, however, the measure of land to be obtained far exceeded the financial limit of the Board. The issue required another methodology, one which was to originate from the rising study of microbiology (Lens & Kuenen, 2001).

The chief scientist working for the Board around then was William Joseph Dibdin (Figure 1.6). Dibdin, a self-taught child of a portrait painter, started work with the Board in 1877, ascending to chief scientific expert in 1882 yet with the obligations of the chief specialist. In looking for an answer for the wastewater disposal issue at the Barking Creek and Crossness outfalls, he started a progression of analyses utilizing different flocculating synthetic chemicals, for example, lime, alum, and ferric chloride—to precipitate the solids before releasing to the stream.

This was not new, obviously, yet Dibdin found that utilizing just a little lime and alum was similarly as powerful as utilizing a ton, and end that spoke to the cost-cognizant Metropolitan Board.

Dibdin perceived that the precipitation procedure did not expel the interest for oxygen, and he had evidently been persuaded by one of his staff, a physicist named August Dupré, that it was important to keep up positive oxygen levels in the water to forestall smells. Dibdin chose to include permanganate of soda (sodium permanganate) to the water to recharge the oxygen levels. Since Dibdin's suggestions were significantly more affordable than the other options, the Board obliged his plan (Van Loosdrecht et al., 1997).

Figure 1.6: William Dibdin (1850–1925).

[Source: https://www.gracesguide.co.uk/William_Joseph_Dibdin]

Dibdin's plan was taken, and in 1885 development of the sewage treatment works at the Barking outfall started. Given the degree of misconception at the time, there were a large number of people who questioned that Dibdin's plan would work, so he needed to persistently defend his work. He again contended that the existence of the addition of the permanganate of soda was important to hold the smell down, and he started to clarify this by recommending that it was important to keep the oxygen aerobic microorganisms healthy. Christopher Hamlin, an antiquarian at the University of Notre Dame, has composed generally on Victorian sanitation and accepts this was a legitimization on Dibdin's part and that he did not yet have knowledge into natural treatment. The more Dibdin was tested by his detractors, nonetheless, the more he obviously turned into a promoter of beneficial aerobic microbiological action in the water since this was his one really one of a kind commitment that could not be invalidated.

When Dibdin began to direct his analyses at the outfall, Dupré was leading trials with aerobic microorganisms. Dupré, a German émigré, and a general health scientific expert contended that moment aerobic microorganisms washed down waterways, and, consequently, these equivalent microorganisms may be utilized for treating wastewater. Dupré attempted to change over Dibdin to the comprehension of microbial activity. In one letter to him, Dupré stated, "The destruction of organic matter discharged into the river in the sewage is practically wholly accomplished by minute organisms. These organisms, however, can only do their work in the presence of oxygen, and the more of that you supply the more rapid the destruction." Later, in an 1888 speech to the Royal Society of Arts, Dupré proposed that "our treatment should be such as to avoid the killing of these organisms or even hampering them in their actions, but rather to do everything to favor them in their beneficial work."

Yet Dupré and Dibdin were not absolutely effective in persuading the Board that their thoughts were correct. Numerous researchers, as yet considering the harms of the microbial world, contended that smell control could be accomplished uniquely by executing the microorganisms. These researchers oversaw in 1887 to wrest control of the treatment works from Dibdin and started a late spring freshening up control recommended by a school educator that included clean treatment with chloride of lime and sulfuric corrosive. This procedure was not successful; Dibdin was vindicated, and biological wastewater treatment turned into the standard for all huge city sewage offices.

1.3.6. The oxygen aerobic microorganisms healthy

Attention to metropolitan solid waste issues was extraordinarily elevated by the saga of the trash barge (Perlman, 1998). It was year 1987, when the barge named Mobro had been stacked in New York with city solid waste and ended up with no place to release the heap. Since disposal into the sea is illicit, the barge was pulled from port to port, with six states and three nations dismissing the chief's pleas to offload its undesirable load (Figure 1.7).

The media got on this unfortunate occurrence and trumpeted the alleged trash crisis to any individual who might tune in. Correspondents sharpened their finest metaphor, asserting that the barge could not empty since all our landfills were full and that the US would before long be covered by solid waste across the nation. Except if we accomplished something soon, they asserted, we could all be choked in the trash (Feinstein, 1989; Bennett & Lawrence, 1995).

The tale of the hapless Mobro is really a story of a pioneering undertaking turned sour. An Alabama specialist, Lowell Harrelson, needed to build an office for changing over metropolitan waste to methane gas and perceived that baled reject would be the best type of denying for that reason. He obtained the bales of metropolitan solid waste from New York City and was going to find a landfill someplace on the east coast or in the Caribbean where he could store the bales and start making methane. Shockingly, he did not get the best possible grants for bringing deny into different districts, and the barge was denied authorization to offload its load. As the voyage proceeded, the press inclusion developed, and no nearby legislators would consent to enable the trash to enter their ports. Poor Harrelson finally needed to consume his interest in a Brooklyn incinerator (Simmons & Widmar, 1989; Melosi, 1993).

Figure 1.7: The garbage barge.

[Source: https://www.nydailynews.com/new-york/trash-fight-long-voyage-new-york-unwanted-garbage-barge-article-1.812895]

The trash crisis never created, obviously. Huge waste disposal companies constructed gigantic landfills in remote regions and started going after the metropolitan solid waste from the eastern seaboard. Though, business visionaries keep on looking for approaches to extricate an incentive from the trash (Clapp, 2002; Roberts, 2016).

1.4. SUSTAINABILITY AND CRADLE-TO-CRADLE DESIGN

The success of the Western world can be viewed as to a great extent a result of the Industrial Upheaval. While the industrialization of the previous two centuries created huge benefits, it likewise left us with a heritage of unintended and negative outcomes: huge amounts of waste, the consumption of natural resources, and the sullying of individuals and biological systems with dangerous substances scattered all through the planet (Lee & Bony, 2008; Bakker et al., 2010).

The Revolution of Industry is in part dependent on a support to-grave model with an example of "take, make, and waste." Items go from being crude materials to finished products and afterward to waste quite promptly. While an item itself might be expected for quick utilization, it might create waste that stays around for many years (Stegall, 2006; Toxopeus et al., 2015). For instance, some food products need to remain fresh for only a couple of days, yet they are bundled in materials that could take many years to decay by common procedures in the atmosphere. In the previous decades, researchers, specialists, and policymakers have started to address the deficiencies of this unfeasible model and to define and try different things with what might be alluring (McDonough & Braungart, 2002; 2003).

Numerous individuals have sought nature for motivation and models of progressively reasonable approaches to make and oversee synthetic chemicals, materials, and items in the public eye. Sustainable materials management (SMM) has risen as "a way to deal with advance sustainable materials use, coordinating activities focused at lessening negative ecological effects and preserving common capital for the duration of the life-cycle of materials, considering financial efficiency and social equity" (Vesilind & Morgan, 2004; Barnes et al., 2009).

1.5. FRAMEWORK FOR SUSTAINABILITY

In 1989, a system for manageability called The Natural Step rose up out of Sweden through the endeavors of Dr. Karl-Henrik Robèrt, a main Swedish oncologist (Glavič & Lukman, 2007). This edge work gives a lot of four framework conditions that define an economical society dependent on the laws of thermodynamics and natural cycles. The Natural Step System Condition considers the Earth as a closed framework for materials and as an open framework for vitality that supports life through a complex intelligent system of material cycles that utilizes solar energy to balance the inclination of materials to scatter and generally increment in entropy (Waggoner & Ausubel, 2002; Kajikawa, 2008).

Consequently, for a society to be maintainable, nature must not be endangered to the following systematically growing procedures:

1. *Extracting convergences of substances from the crust of the Earth.* This condition alludes to the extraction of minerals and non-renewable energy sources. Substances that are rare in nature ought to be subbed with those that are increasingly inexhaustible. Mined materials ought to be utilized efficiently and reused, and reliance on non-renewable energy sources ought to be methodologically diminished.

Figure 1.8: Ecological aspects of The Natural Step System Conditions. (Illustrated by Larry Chalfant and Lauren Heine)

[Source: https://www.amazon.com/Introduction-Environmental-Engineering-Aarne-Vesilind/dp/0495295833]

2. *Building up concentrations of human-made mixes in nature.* This condition alludes to the assembling of persevering and unnatural mixes. Persistent and unnatural mixes ought to be supplanted with those that are ordinarily abundant or potentially that separate totally and effectively in nature. All substances delivered by society ought to be utilized efficiently.

3. *Utilizing renewable assets at rates quicker than they are recovered and decreasing the beneficial limit of nature.* This condition alludes to the utilization of natural assets. Assets ought to be drawn uniquely from all around oversaw ecosystems, efficiently seeking after the most beneficial and manageable uses both of

those assets and land, and practicing alert in a wide range of modification of nature.

And in that society:

4. *People can address their issues around the world.* This condition implies utilizing the entirety of our assets efficiently, viably, decently and mindfully with the goal that the necessities surprisingly, including the future needs of individuals who are not yet conceived, stand the most obvious opportunity with regards to being met.

The environmental parts of The Natural Step System Conditions are outlined in Figure 1.8. Materials flow in a closed framework contained two circles. The external circle speaks about the cycling of materials inside the earth's biological systems (O'Connor, 2006; Sala et al., 2015). The internal circle speaks about cycling inside the modern/monetary framework. Arrow 1 shows the extraction of natural assets for use in the mechanical/financial system. In a manageable society, the pace of characteristic asset extraction rises to the pace of recovery (Cai et al, 2002; Beach et al., 2009). Arrows 2 and 3 display the extraction and resettling of materials from the crust of the Earth, principally non-renewable energy source and mined materials. In a manageable society, material extraction from the world's covering will be dislodged by the utilization of reused and recyclable materials. Arrows 4 and 5 show sub-positions that flow from the mechanical/monetary framework to the more prominent environment. Substances that acclimatize rapidly without harm are spoken to by Bolt 4. Bolt 5 speaks to sub-positions that are lethal, as shown, bioaccumulative, or generally cause damage to people or nature. In a supportable society, Arrow 5 will vanish (Anderies et al., 2007; Heijungs et al., 2010).

1.6. CRADLE-TO-CRADLE DESIGN

One of the main feasible methodologies for accomplishing sustainable materials the board is through positive industrial action, called "support to-support" structure. The expression "support to-support was begotten during the 1970s by Michael Braungart and William McDonough (2002). The key roles of support to-support configuration were first efficiently delineated as the Intelligent Product System (IPS) by Braugart et al. (1992) further created and explained by Michael Braungart and William McDonough in 2002 in their book Cradle to Cradle: Remaking the Way We Make Things. Similarly as in natural frameworks where one life form's waste becomes nourishment for another, cradle-to-cradle configuration applies a similar idea to the plan of human industry. Cradle-to-cradle plan defines two digestion systems inside which materials are imagined as supplements circling considerately and profitably through digestion systems. Biological supplements cycle inside natural digestion systems, and specialized supplements cycle inside specialized digestion systems (Braungart et al., 2007).

Organic digestion is the arrangement of normal procedures that supports life. Natural procedures are repeating, at last powered by the vitality of the sun, and incorporate the bio-degradation (and potentially different types of degradation) of natural materials and their consolidation into life forms. Materials that add to the efficiency of natural digestion systems are organic supplements. They are sustainable, degradable, and biologically kind. Results of industry produced using organic supplements can be incorporated into natural or designed natural digestion systems, including water treatment procedures and natural processing systems, for example, fertilizing the soil or anaerobic assimilation. The yield of organic digestion systems can be assets that incite new natural supplements, for example, beneficial soil alterations. Items that are expected for discharge to the earth should structure as organic supplements that are kindhearted for their planned utilitarian use (Rossi et al., 2006).

The industry can likewise imitate natural processes by making specialized digestion systems that circulate specialized supplements. Specialized supplements are mostly nonrenewable and they are important for their performance characteristics. Models incorporate metals, for example, aluminum or copper. At the point when structured in cradle-to-cradle frameworks, specialized supplements can be recouped and reused again and again—without debasing their quality and without damage to handlers—into comparable or divergent items. A few organizations see the materials in their items as so significant that they even participate in renting programs whereby items are basically rented from the producer to the client until they are never again needed. At that point, the maker will take them back for remanufacturing of the important materials and parts into new items. Specialized supplements can be intended for reuse inside an organization or between organizations incomparable or divergent ventures, contingent upon the material. Items produced using specialized supplements ought to be intended to encourage material recovery at its most noteworthy incentive with negligible consumption of vitality and cost. Cradle-to-cradle design, as described by McDonough and Braungart, uses a model of human industry based on three design principles derived from natural systems (Tanzil et al., 2005).

1. *Use the current sun based salary.* With not very many special cases, life on earth is at last energized by vitality from the sun. We are just starting to grow our ability to outfit sun oriented energy, legitimately and in a roundabout way, for human purposes.
2. *Celebrate assorted variety.* Natural frameworks flourish with lavishness and decent variety. In like manner, the industry ought to advance the improvement of assorted items that are fitting for various inclinations, societies, topographies, and environments.
3. *Waste equivalents nourishment.* There is no loss in nature. The result of one creature is nourishment or structure for another. Human frameworks can likewise be intended to course materials beneficially, dispensing with the idea of waste.

A few people require a methodology of eco-efficiency: to lessen the measure of resources utilized and to produce less waste in industrial work. Be that as it may, eco-efficiency alone is definitely not a sufficient methodology for sustainability. Upgrades in eco-efficiency are frequently immediately overpowered by increments sought after. For instance, improvement in vehicle fuel efficiency has been balanced by an expansion in the number of autos on the streets and the number of miles driven. Consider eco-efficiency as an approach to slow the loss of assets, like helping an injury drain gradually and not hemorrhage. Cradle-to-cradle configuration calls for eco-viability, which is closely resembling mending the injury and supporting the soundness of the entire body. Eco-adequacy is worried about expanding repeating material flows (subsequently "viability"), so waste approaches nourishment.

By updating industry dependent on nature's models, monetary movement can fortify, as opposed to bargain, social and ecological flourishing. While this may sound somewhat like pure fantasy, there are some unintended benefits of utilizing support to-support as a structure technique. Most importantly, it drives development. Organizations are continually searching for new thoughts and approaches to recognize their organizations, to increase the value of their items, and, obviously, to make more profit. Support to-support configuration invigorates crisp reasoning and creativity. According to Roger McFadden, presently boss science officer and VP for item science and innovation for Corporate Express a Staples Organization, "Chemical product manufacturers should recognize the opportunity that the sustainability movement is creating for innovation. The movement is exciting in part because it is driving real product innovation and development of new raw materials for formulation after a long period of incremental change" (O'riordan, 1986; Kozarev & Ilieva, 2011).

What are the benefits of items intended for support to-support frameworks? For one, the general population is in an inexorably green mindset! Allowed the chance to purchase two items with comparable execution and comparable value, who might not want to purchase the one with greener synthetic chemicals and inventive bundling that dispenses with waste? Support to-support configuration likewise drives imagination around new plans of action. For instance, items structured as specialized supplements can be seen as results of administration. Looking at the situation objectively, do individuals truly need to own PCs or TVs? (Presently you are certain we are dreaming, right?) What individuals truly need is to have the option to bear the cost of and select the most excellent hardware feasible for their unhindered satisfaction in the protection of their homes (or offices). In the event that when you were prepared to change your model, you realized that you could helpfully restore the item to the producer and that its components would be reused or revamped, OK mind? We foresee not, particularly if its arrival brought you to value, for example, discounts or limits toward the following thing of administration that you want (Korobitsina et al., 1969).

Support to-support thinking may come as natural to ecological architects, who are accused of ensuring both human health and the earth and who figure out how to structure material flow frameworks for materials, for example, water and wastewater. Instead of treating the loss toward the end of the funnel, ecological architects can utilize the instruments of the exchange to plan economical frameworks in advance that reuse important resources without debasing their quality after some time. While we cannot prevent the present existence from the foundation, it should not stop us from improving plans and applying support to-support thinking as these frameworks age and should be supplanted and as new frameworks are required. It merits thinking about the structures (and plan disappointments) of the past as models for better and better structures. As shown by the items depicted in the containers in this section, the structure of the item decides the finish of-life material choices. Specialists can influence the adequacy of the framework by comprehension and influencing the item plan and by not considering one without the other.

1.6.1. Herman Miller Mirra Office Chair

The Herman Mill operator Mirra office seat was the first office seat intended to flow in specialized cycles as indicated by the standards of support to-support structure. The seat was the consequence of joint effort between Herman Mill operator, Studio 7.5, McDonough Braungart Plan Science and EPEA Internationale Umweltforschung GmbH. The objective was to synchronize item improvement with support to-support plan (Goeman & Beckwith, 2004; Shipper et al., 2014).

The group made a Design for the Environment (DfE) item appraisal instrument to assess the ecological exhibition of new item plans in three key zones: material science, dismantling, and recyclability. Assessing material science includes three center advances:

1. Identifying every one of the synthetic substances in a material used to fabricate an item, for example, the steel shaft in a chair—down to 100 ppm.
2. Evaluating the risks presented by the synthetic compounds in the material.
3. Assigning the material a score of green, yellow, orange, or red dependent on green being practically zero danger yellow being low to direct risk orange being inadequate information red being high hazard.

Dismantling alludes to the simplicity of breaking a final item, for example, the office chair—down into its constituent parts for reusing or reuse. Assessing recyclability thinks about whether a section contains reused material and, all the more critically, regardless of whether that part can be reused toward the finish of the item's valuable life (Unruh, 2008).

The Mirra chair was intended for specialized cycles as a result of the administration. Much the same as supplements in the natural world, materials in specialized cycles are supplements in the mechanical world and can be utilized on various occasions. The materials utilized in the Mirra chair were broke down by EPEA in Hamburg, Germany. Materials that are dangerous or incapable of flow in a specialized cycle were disposed of from the structure. Of the synthetic compounds and materials chose, 69% of them are viewed as green. Regarding reusing, the seat contains 42% by weight pre-and postconsumer reused content, and 96% by weight of the chair is recyclable. As for dismantling, 93% of the item by weight can be promptly dismantled, a need to encourage reusing (Figure 1.9).

Figure 1.9: The Herman Miller Mirra office chair. Left: recyclable parts, 96% by weight. Right: non-recyclable parts (4% by weight)—mixed plastic arm pads (white parts), seat pan, and leaf springs (black parts). © Courtesy Hermann Miller Canada, Inc.

[Source: https://www.amazon.com/Introduction-Environmental-Engineering-Aarne-Vesilind/dp/0495295833]

1.6.2. Eco-Effective Chemistry

Baypure® is a line of business synthetic compounds developed to have high natural execution while meeting or surpassing the performance of customary synthetics. It comprises of synthetic substances utilized in mechanical procedures that were intended to have improved biodegradability, great eco-toxicological qualities, and create forms with closed circle material flows (Gungor & Evans, 2015; Teng et al., 2016). The Baypure® line is made up essentially of two synthetic substances: Baypure® DS100 is a scattering specialist and Baypure CX100 is a complex operator. Scattering operators help dissolve solids in fluids and are utilized, for instance, to scatter strong shades into fluids during the production of synthetic colors. A complex operator ties metals in fluid arrangements, preventing them from meddling adversely with item execution. They are utilized, for instance, in cleansers and in a paper generation (Braungart et al., 2007; Bai et al., 2015).

EPEA Internationale Umweltforschung GmbH thought about the ecological exhibition of Baypure® DS100 and CX100 to their conventional partners over the whole life-cycle of the synthetic substances, from crude material procurement to generation procedures to their destiny in the indigenous habitat after use. The focal test was to guarantee not just that the substance performs well for its ideal capacity yet that it supports the natural condition all through the item life cycle. EPEA found that, compared with their normal counterparts, the Baypure® line of synthetic concoctions exceeds expectations in their conduct in the earth, especially after use and release into the water. For instance, the traditional partner of Baypure® CX100 is EDTA. EDTA is a generally utilized complexing operator that degrades gradually in the normal environment, prepares metals, and is risky in the amphibian condition. Conversely, CX100 has low poisonous quality and is biodegradable, making it compatible with the natural environment (López & Montalvo, 2015).

REFERENCES

1. Abbott, N. L. (2001). New horizons for surfactant science in chemical engineering. *American Institute of Chemical Engineers. AIChE Journal*, *47*(12), 2634.
2. Abraham, M. A. (2006a). Principles of sustainable engineering. In *Sustainability Science and Engineering* (Vol. 1, pp. 3–10). Elsevier.
3. Abraham, M. A. (2006b). Sustainability: Philosophy vs. engineering tools. Environmental Progress, 25(2), 87–88.
4. Abraham, M. A. (Ed.). (2005). Sustainability Science and Engineering: Defining Principles (Vol. 1, pp. 11–20). Elsevier.
5. Abraham, M. A., & Nguyen, N. (2003). "Green engineering: Defining the principles"—results from the Sandestin conference. *Environmental Progress*, *22*(4), 233–236.
6. Amdur, M. O., & Corn, M. (1963). The irritant potency of zinc ammonium sulfate of different particle sizes. *American Industrial Hygiene Association Journal*, *24*(4), 326–333.
7. An, J., Zhang, M., Wang, S., & Tang, J. (2008). Physical, chemical and microbiological changes in stored green asparagus spears as affected by coating of silver nanoparticles-PVP. *LWT-Food Science and Technology*, *41*(6), 1100–1107.
8. ANASTAS, P. (2003). Design through the 12 principles of green engineering. *Environ. Sci. Tech.*, *37*, 94A–101A.
9. Anderies, J. M., Rodriguez, A. A., Janssen, M. A., & Cifdaloz, O. (2007). Panaceas, uncertainty, and the robust control framework in sustainability science. *Proceedings of the National Academy of Sciences*, *104*(39), 15194–15199.
10. Anderson, J. L., Chenoweth, S., DeVasher, R., House, R., Livingston, J., Minster, M., ... & Williams, J. M. (2008, July). Communicating sustainability: Sustainability and communication in the engineering, science, and technical communication classrooms. In *2008 IEEE International Professional Communication Conference* (Vol.1, pp. 1–7). IEEE.
11. Bai, R., Liu, P., Yang, J., Liu, C., & Gu, Y. (2015). Facile Synthesis of 2-Aminothiophenes Using NaAlO2 as an Eco-Effective and Recyclable Catalyst. *ACS Sustainable Chemistry & Engineering*, *3*(7), 1292–1297.
12. Bakker, C. A., Wever, R., Teoh, C., & De Clercq, S. (2010). Designing cradle-to-cradle products: a reality check. *International Journal of*

Sustainable Engineering, *3*(1), 2–8.

13. Barnes, B., Cao, H. T., Drab, T., & Pearson, J. (2009). Design of sustainable relief housing in Ethiopia: an implementation of cradle to cradle design in earthbag construction. *American Journal of Environmental Sciences*, *5*(2), 137–144.
14. Bartlett, L., & Vesilind, P. A. (1998). Expediency and human health: The regulation of environmental chromium. *Science and Engineering Ethics*, *4*(2), 191–201.
15. Beach, E. S., Cui, Z., & Anastas, P. T. (2009). Green Chemistry: A design framework for sustainability. *Energy & Environmental Science*, *2*(10), 1038–1049.
16. Bennett, W. L., & Lawrence, R. G. (1995). News icons and the mainstreaming of social change. *Journal of Communication*, *45*(3), 20–39.
17. Bielefeldt, A. (2013). Pedagogies to achieve sustainability learning outcomes in civil and environmental engineering students. *Sustainability*, *5*(10), 4479–4501.
18. Bielefeldt, A. R. (2011). Incorporating a sustainability module into first-year courses for civil and environmental engineering students. *Journal of Professional Issues in Engineering Education and Practice*, *137*(2), 78–85.
19. Braungart, M., McDonough, W., & Bollinger, A. (2007). Cradle-to-cradle design: creating healthy emissions–a strategy for eco-effective product and system design. *Journal of Cleaner Production*, *15*(13–14), 1337–1348.
20. Braungart, M., McDonough, W., & Bollinger, A. (2007). Cradle-to-cradle design: creating healthy emissions–a strategy for eco-effective product and system design. *Journal of Cleaner Production*, *15*(13–14), 1337–1348.
21. Bryan, J. A., Lehmann, J. D., Setiady, I. F., & Hatch, M. H. (1974). An outbreak of hepatitis: A associated with recreational lake water. *American Journal of Epidemiology*, *99*(2), 145–154.
22. Burke, T., Fagliano, J., Goldoft, M., Hazen, R. E., Iglewicz, R., & McKee, T. (1991). Chromite ore processing residue in Hudson County, New Jersey. *Environmental Health Perspectives*, *92*, 131–137.
23. Cai, X., McKinney, D. C., & Lasdon, L. S. (2002). A framework for sustainability analysis in water resources management and application

to the Syr Darya Basin. *Water Resources Research, 38*(6), 21–1.

24. Chadwick, A., & Morfett, J. (2002). *Hydraulics in Civil and Environmental Engineering (Vol. 1, pp.11–30)*. CRC Press.
25. Chang, L., & O'Brien, T. (1970). Australia antigen serology in the Holy Cross football team hepatitis outbreak. *The Lancet, 296*(7663), 59–61.
26. Ciocco, A., & Thompson, D. J. (1961). A follow-up of Donora ten years after methodology and findings. *American Journal of Public Health and the Nations Health, 51*(2), 155–164.
27. Clapp, J. (2002). The distancing of waste: Overconsumption in a global economy. *Confronting Consumption*, 1, 155–176.
28. Cooke, S. J., & Vermaire, J. C. (2015). Environmental studies and environmental science today: inevitable mission creep and integration in action-oriented transdisciplinary areas of inquiry, training and practice. *Journal of Environmental Studies and Sciences, 5*(1), 70–78.
29. Craun, G. F., McCabe, L. J., & Hughes, J. M. (1976). Waterborne disease outbreaks in the US—1971–1974. *Journal-American Water Works Association, 68*(8), 420–424.
30. Crusberg, T. C., Burke, W. M., Reynolds, J. T., Morse, L. E., Reilly, J., & Hoffman, A. H. (1978). The reappearance of a classical epidemic of infectious hepatitis in Worcester, Massachusetts. *American Journal of Epidemiology, 107*(6), 545–551.
31. Daims, H., Taylor, M. W., & Wagner, M. (2006). Wastewater treatment: a model system for microbial ecology. *Trends in Biotechnology, 24*(11), 483–489.
32. Davis, R. D., & Hall, J. E. (1997). Production, treatment and disposal of wastewater sludge in Europe from a UK perspective. *European Water Pollution Control, 2*(7), 9–17.
33. Deng, G. Z., Wang, X. D., & Che, X. K. (2003). Today and Tomorrow of Titanium Industry [J]. *Iron Steel Vanadium Titanium, 1, 1–30.*
34. Epstein, E., Willson, G. B., Burge, W. D., Mullen, D. C., & Enkiri, N. K. (1976). A forced aeration system for composting wastewater sludge. *Journal (Water Pollution Control Federation)*,1, 688–694.
35. Feinstein, A. R. (1989). Para-analysis, faute de mieux, and the perils of riding on a data barge. *Journal of Clinical Epidemiology, 42*(10), 929–935.
36. Field, A. (1973). Environmental engineering. In *Mechanical Engineer's Reference Book* (Vol.1, pp. 15–24). Butterworth-Heinemann.

37. Forthman, D. L., & Ogden, J. J. (1992). The role of applied behavior analysis in zoo management: today and tomorrow. *Journal of Applied Behavior Analysis*, *25*(3), 647.
38. Freema, N. C., Stern, A. H., & Lioy, P. J. (1997). Exposure to chromium dust from homes in a chromium surveillance project. *Archives of Environmental Health: An International Journal*, *52*(3), 213–219.
39. Freeman, N. C. G., Wainman, T., Lioy, P. J., Stern, A. H., & Shupack, S. I. (1995). The effect of remediation of chromium waste sites on chromium levels in urine of children living in the surrounding neighborhood. *Journal of the Air & Waste Management Association*, *45*(8), 604–614.
40. Freeman, N. C., Lioy, P. J., & Stern, A. H. (2000). Reduction in residential chromium following site remediation. *Journal of the Air & Waste Management Association*, *50*(6), 948–953.
41. Friedman, L. S., O'Brien, T. F., Morse, L. J., Chang, L. W., Wacker, W. E., Ryan, D. M., & Dienstag, J. L. (1985). Revisiting the Holy Cross football team hepatitis outbreak (1969) by serological analysis. *JAMA*, *254*(6), 774–776.
42. García-Serna, J., Pérez-Barrigón, L., & Cocero, M. J. (2007). New trends for design towards sustainability in chemical engineering: Green engineering. *Chemical Engineering Journal*, *133*(1–3), 7–30.
43. Gerogiorgis, D. I., & Jolliffe, H. G. (2015). Continuous pharmaceutical process engineering and economics. *Chim. Oggi-Chem. Today*, *33*(6), 29–32.
44. Glavič, P., & Lukman, R. (2007). Review of sustainability terms and their definitions. *Journal of Cleaner Production*, *15*(18), 1875–1885.
45. Goeman, D., & Beckwith, D. (2004). Design's strategic role at Herman Miller. *Design Management Review*, *15*(2), 40–45.
46. Grose, T. K. (2006). Trouble on the Horizon. *ASEE Prism*, *16*(2), 27.
47. Gungor, Z. E., & Evans, S. (2015). Eco-effective changeovers; changing a burden into a manufacturing capability. *Procedia CIRP*, *26*, 527–532.
48. Heijungs, R., Huppes, G., & Guinée, J. B. (2010). Life cycle assessment and sustainability analysis of products, materials and technologies. Toward a scientific framework for sustainability life cycle analysis. *Polymer Degradation and Stability*, *95*(3), 422–428.
49. HG, I. (2001). Assisting nature: William Dibdin and biological wastewater treatment. *Water Resources Impact, 1, 1–12.*

50. Hoffman, A. H., Crusberg, T. C., & Sayilonis, B. J. (1979). Viral hepatitis and hydraulic parameters: an alternative hypothesis. *Archives of Environmental Health: An International Journal*, *34*(2), 87–91.
51. Hossain, M. K., Strezov, V., & Nelson, P. F. (2009). Thermal characterization of the products of wastewater sludge pyrolysis. *Journal of Analytical and Applied Pyrolysis*, *85*(1–2), 442–446.
52. Jingjiao, Z. (1988). Genesis and Research Significance of the Plinthitic Horizon [J]. *Geographical Research, 4, 1–30.*
53. Jones, D. A., Lelyveld, T. P., Mavrofidis, S. D., Kingman, S. W., & Miles, N. J. (2002). Microwave heating applications in environmental engineering—a review. *Resources, Conservation and Recycling*, *34*(2), 75–90.
54. Junwei, L., Tingzhou, L., Shuhua, Y., Zaifeng, L., & Xiaofeng, H. (2012). Garbage-fired power generation in china today and development trends. *Sino-Global Energy*, *6,* 1–20.
55. Kajikawa, Y. (2008). Research core and framework of sustainability science. *Sustainability Science*, *3*(2), 215–239.
56. Kelessidis, A., & Stasinakis, A. S. (2012). Comparative study of the methods used for treatment and final disposal of sewage sludge in European countries. *Waste Management*, *32*(6), 1186–1195.
57. Kitsa, V., Lioy, P. J., Chow, J. C., Watson, J. G., Shupack, S., Howell, T., & Sanders, P. (1992). Particle-size distribution of chromium: total and hexavalent chromium in inspirable, thoracic, and respirable soil particles from contaminated sites in New Jersey. *Aerosol Science and Technology*, *17*(3), 213–229.
58. Klopfer, E., & Yoon, S. (2004). Developing games and simulations for today and tomorrow's tech-savvy youth. *TechTrends*, *49*(3), 33–41.
59. Korobitsina, I. K., Gurevich, L. S., & Rodina, L. L. (1969). Chemistry of Ketones of Furanidine Series. IX. Alkaline Cleavage of 2, 2, 5, 5-Tetraalkyl-3, 4-furanidindiones. *Russ. J. Org. Chem*, *5*, 567–570.
60. Korshover, J., & Angell, J. K. (1982). A review of air-stagnation cases in the eastern United States during 1981—Annual summary. *Monthly Weather Review*, *110*(10), 1515–1518.
61. Kozarev, N., & Ilieva, N. (2011). Plume rise in particular meteorological conditions. *Journal of the University of Chemical Technology and Metallurgy*, *46*(3), 305–308.
62. LaPara, T. M., Nakatsu, C. H., Pantea, L. M., & Alleman, J. E. (2002).

Stability of the bacterial communities supported by a seven-stage biological process treating pharmaceutical wastewater as revealed by PCR-DGGE. *Water Research, 36*(3), 638–646.

63. Leal Filho, W. (2012). *Sustainable Development at Universities: New Horizons. Umweltbildung, Umweltkommunikation und Nachhaltigkeit/ Environmental Education, Communication and Sustainability. Volume 34, pp. 40–55.* Frankfurt am Main, Germany.
64. Lee, D., & Bony, L. J. (2008). Cradle-to-cradle design at Herman Miller: moving toward environmental sustainability. *HBS Case*, (Vol. 607–003, pp. 10–40).
65. Lens, P. N. L., & Kuenen, J. G. (2001). The biological sulfur cycle: novel opportunities for environmental biotechnology. *Water Science and Technology*, *44*(8), 57–66.
66. Lioy, P. J., Freeman, N. C. G., Wainman, T., Stern, A. H., Boesch, R., Howell, T., & Shupack, S. I. (1992). Microenvironmental analysis of residential exposure to chromium-laden wastes in and around New Jersey homes. *Risk Analysis*, *12*(2), 287–299.
67. Liu, Y. (2003). Chemically reduced excess sludge production in the activated sludge process. *Chemosphere*, *50*(1), 1–7.
68. López, F. J. D., & Montalvo, C. (2015). A comprehensive review of the evolving and cumulative nature of eco-innovation in the chemical industry. *Journal of Cleaner Production*, *102*, 30–43.
69. Lu, H., Chandran, K., & Stensel, D. (2014). Microbial ecology of denitrification in biological wastewater treatment. *Water Research*, *64*, 237–254.
70. Luthy, R., & Small, M. (1990). ES&T Views: Environmental research: A clearer focus over a broader horizon. *Environmental Science & Technology*, *24*(11), 1620–1623.
71. Masuda, S. (1988). Pulse corona induced plasma chemical process: a horizon of new plasma chemical technologies. *Pure and Applied Chemistry*, *60*(5), 727–731.
72. McDonough, W., & Braungart, M. (2002). Design for the triple top line: new tools for sustainable commerce. *Corporate Environmental Strategy*, *9*(3), 251–258.
73. McDonough, W., & Braungart, M. (2002). *Cradle to Cradle: Remaking the Way We Make Things* (No. 504.064. 4, pp. 11–30). North Point Press.

74. McDonough, W., & Braungart, M. (2003). Towards a sustaining architecture for the 21st century: the promise of cradle-to-cradle design. *Industry and Environment*, *26*(2), 13–16.
75. Melosi, M. V. (1993). Down in the dumps: Is there a garbage crisis in America?. *Journal of Policy History*, *5*(1), 100–127.
76. Mencken, H. L. (1945). *The American Language: An Inquiry Into the Development of English in the United States* (Vol. 2). Alfred a Knopf Incorporated.
77. Meyers, J. D., Huff, J. C., Holmes, K. K., Thomas, E. D., & Bryan, J. A. (1974). Parenterally transmitted hepatitis A associated with platelet transfusions: epidemiologic study of an outbreak in a marrow transplantation center. *Annals of Internal Medicine*, *81*(2), 145–151.
78. Morse, L. J., Bryan, J. A., Hurley, J. P., Murphy, J. F., O'Brien, T. F., & Wacker, W. E. (1972). The Holy Cross College football team hepatitis outbreak. *Jama*, *219*(6), 706–708.
79. Narayanasamy, S., Muller, E. E., Sheik, A. R., & Wilmes, P. (2015). Integrated omics for the identification of key functionalities in biological wastewater treatment microbial communities. *Microbial Biotechnology*, *8*(3), 363–368.
80. North, E. W., Adams, E. E., Schlag, Z., Sherwood, C. R., He, R., Hyun, K. H., & Socolofsky, S. A. (2011). Simulating oil droplet dispersal from the Deepwater Horizon spill with a Lagrangian approach. *Geophys. Monogr. Ser*, *195*, 217–226.
81. O'Connor, M. (2006). The "Four Spheres" framework for sustainability. *Ecological Complexity*, *3*(4), 285–292.
82. Ødegaard, H., Paulsrud, B., & Karlsson, I. (2002). Wastewater sludge as a resource: sludge disposal strategies and corresponding treatment technologies aimed at sustainable handling of wastewater sludge. *Water Science and Technology*, *46*(10), 295–303.
83. O'Riordan, T. (1986). Discussion: Advocating the Rights and Interests of Future Generations. *Schnaiberg/Watts/Zimmermann, eds.(1986)*, 1, 419–433.
84. Ottosen, L. M., Christensen, I. V., Rörig-Dalgaard, I., Jensen, P. E., & Hansen, H. K. (2008). Utilization of electromigration in civil and environmental engineering—processes, transport rates and matrix changes. *Journal of Environmental Science and Health Part A*, *43*(8), 795–809.

85. Paredes, D., Kuschk, P., Mbwette, T. S. A., Stange, F., Müller, R. A., & Köser, H. (2007). New aspects of microbial nitrogen transformations in the context of wastewater treatment: a review. *Engineering in Life Sciences*, *7*(1), 13–25.

86. Perlman, S. (1998). *Barging into a Trashy Saga, 1*, 1–22.

87. Perry, R. H., & Perry, J. H. (1976). *Engineering Manual: A Practical Reference of Design Methods and Data in Building Systems, Chemical, Civil, Electrical, Mechanical, and Environmental Engineering and Energy Conversion* (Vol. 1, pp.11–40). McGraw-Hill.

88. Phelps, E. B. (1948). Public Health Engineering. A Textbook of the Principles of Environmental Sanitation. *Public Health Engineering. A Textbook of the Principles of Environmental Sanitation, 1,* 11–22.

89. Roberts, S. T. (2016). Digital refuse: Canadian garbage, commercial content moderation and the global circulation of social media's waste. *Wi: Journal of Mobile Media, 1*(1), 1–30.

90. Rossi, M., Charon, S., Wing, G., & Ewell, J. (2006). Design for the next generation: incorporating cradle-to-cradle design into Herman Miller products. *Journal of Industrial Ecology*, *10*(4), 193–210.

91. Sala, S., Ciuffo, B., & Nijkamp, P. (2015). A systemic framework for sustainability assessment. *Ecological Economics*, *119*, 314–325.

92. Shipper, F., Manz, K. P., Adams, S. B., & Manz, C. C. (2014). Herman Miller: Unrelenting Pursuit of Reinvention and Renewal. In *Shared Entrepreneurship* (Vol. 1, pp. 133–158). Palgrave Macmillan, New York.

93. Shrenk, H. H., Heimann, H., Clayton, G. D., Gafafer, W. M., & Wexler, H. (1949). Air pollution in Donora, PA: Epidemiology of the unusual smog episode of October 1948. Preliminary report. *Public Health Bulletin*, 1,1, *306*.

94. Simmons, D. A., & Widmar, R. (1989). Participation in household solid waste reduction activities: The need for public education. *Journal of Environmental Systems*, *19*(4), 1–20.

95. Singh, K. P., Mohan, D., Sinha, S., & Dalwani, R. (2004). Impact assessment of treated/untreated wastewater toxicants discharged by sewage treatment plants on health, agricultural, and environmental quality in the wastewater disposal area. *Chemosphere*, *55*(2), 227–255.

96. Snyder, L. P. (1994). "The Death-Dealing Smog over Donora, Pennsylvania": Industrial Air Pollution, Public Health Policy, and the

Politics of Expertise, 1948–1949. *Environmental History Review*, 1, 117–139.

97. Socolofsky, S. A., Adams, E. E., & Sherwood, C. R. (2011). Formation dynamics of subsurface hydrocarbon intrusions following the Deepwater Horizon blowout. *Geophysical Research Letters*, *38*(9), 10–30.
98. Stegall, N. (2006). Designing for sustainability: A philosophy for ecologically intentional design. *Design Issues*, *22*(2), 56–63.
99. Stern, A. H., Freeman, N. C., Pleban, P., Boesch, R. R., Wainman, T., Howell, T., ... & Lioy, P. J. (1992). Residential exposure to chromium waste—urine biological monitoring in conjunction with environmental exposure monitoring. *Environmental Research*, *58*(1–2), 147–162.
100. Tai, H. S., & Jou, C. J. G. (1999). Immobilization of chromium-contaminated soil by means of microwave energy. *Journal of Hazardous Materials*, *65*(3), 267–275.
101. Tanzil, D., Beaver, E. R., Heine, L., & Abraham, M. A. (2005). Designing for Sustainability. *Transforming Sustainability Strategy into Action*; Beloff, B., Lines, M., Tanzil, D., (eds.), *1*, 163–197.
102. Tay, J. H., & Show, K. Y. (1997). Resource recovery of sludge as a building and construction material—a future trend in sludge management. *Water Science and Technology*, *36*(11), 259–266.
103. Taylor, F. B. (1972). The Holy Cross Episode. *Journal American Water Works Association*, *64*(4), 230–232.
104. Teng, J., Ma, H., Wang, F., Wang, L., & Li, X. (2016). A Facile and Eco-Effective Catalytic System for Synthesis of 5-Hydroxymethylfurfural from Glucose. *BioResources*, *11*(1), 2152–2165.
105. Thibodeaux, L. J., Valsaraj, K. T., John, V. T., Papadopoulos, K. D., Pratt, L. R., & Pesika, N. S. (2011). Marine oil fate: knowledge gaps, basic research, and development needs; a perspective based on the Deepwater Horizon spill. *Environmental Engineering Science*, *28*(2), 87–93.
106. Townsend, J. G. (1950). Investigation of the smog incident in Donora, Pa., and vicinity. *American Journal of Public Health and the Nations Health*, *40*(2), 183–189.
107. Toxopeus, M. E., De Koeijer, B. L. A., & Meij, A. G. G. H. (2015). Cradle to cradle: effective vision vs. Efficient practice?. *Procedia CIRP*, *29*, 384–389.

108. Unruh, G. C. (2008). The biosphere rules. *Harvard Business Review, 86*(2), 111–117.
109. Van Loosdrecht, M. C. M., Hooijmans, C. M., Brdjanovic, D., & Heijnen, J. J. (1997). Biological phosphate removal processes. *Applied Microbiology and Biotechnology*, *48*(3), 289–296.
110. Vesilind, P. A., & Morgan, S. M. (2004). Introduction to environmental engineering. *Introduction to Environmental Engineering,* Vol. 2, pp. 1–30.
111. Wacker, W. E., Riordan, J. F., Snodgrass, P. J., Chang, L. W., Morse, L. J., O'Brien, T. F., & Reddy, W. J. (1972). The Holy Cross hepatitis outbreak: Clinical and chemical abnormalities. *Archives of Internal Medicine*, *130*(3), 357–360.
112. Waggoner, P. E., & Ausubel, J. H. (2002). A framework for sustainability science: A renovated IPAT identity. *Proceedings of the National Academy of Sciences*, *99*(12), 7860–7865.
113. Wedeen, R. P., Haque, S., Udasin, I., D'Haese, P. C., Elseviers, M., & De Broe, M. E. (1996). Absence of tubular proteinuria following environmental exposure to chromium. *Archives of Environmental Health: An International Journal*, *51*(4), 321–323.
114. Winkelstein, W. (1972). Utility or futility of ordinary mortality statistics in the study of air pollution effects. In *Proceedings of the Sixth Berkeley Symposium on Mathematical Statistics and Probability, Volume 6: Effects of Pollution on Health* (pp. 1–55). The Regents of the University of California.
115. Witmer, C. M., Harris, R., & Shupack, S. I. (1991). Oral bioavailability of chromium from a specific site. *Environmental Health Perspectives*, *92*, 105–110.
116. Xiao-zhou, L. I. U. (2007). Copper Resources in China: Today and Tomorrow [J]. *Northwestern Geology*, *1*, 1–20.

CHAPTER 2

Global Aspects of Environmental Engineering

CONTENTS

2.1. INTRODUCTION

The global society and supremacy of humans over the biosphere have triggered a situation where environmental and ecological engineering are the unavoidable professional reaction to the most pressing issues of humankind. The aim of both practical and theoretical environmental engineering is to assist society's function (Dresp, 2008). Combining theory, practice, research, and education describes the basis for the altercation of knowledge compared to the engineering solutions that are generally socio-economically justifiable, liable, and consistent with the ecological development of communities, nations, and populations. Such solutions target at designing, planning, and improving structures, infrastructures and facilities for the responsible, economic, and effective utilization of the energy resources (Noguera et al., 1999; Dalby, 2013). For the safeguard of the health of the environment and man, environmental engineering comprises identifying and averting pollutant behaviors in the natural and man-made systems, and, beyond all, to reduce the negative influence, or the ecological footprint, of the humankind on all cycles and systems upon which the life depends (Park et al., 2003; El-Fadel et al., 1997). This chapter offers a brief analysis of historic background facts, theoretical concepts, philosophical considerations, and pragmatic problems from which modern environmental engineering has ascended, as both the scientific discipline and the domain of pragmatic research and development (Jabbour & Santos, 2008).

The influence of human activity on the biosphere has yielded the global social framework in which a shortage of natural resources and dangers to ecological health like water contamination and air pollution call for fresh solutions that aid withstands the development of human society and all of the life on earth (Salic & Zelic, 2018). This chapter starts by recalling the philosophical and historical context from which modern environmental engineering has ascended as the science and domain of technological development. Examples that generally deal with the core challenges and issues presently encountered by the field, like the issues of scale and intricacy, are then discussed. It is highlighted that the sustainability of the made environment is dependent on building designs and innovative architecture for optimal utilization and recycling of the resources (Young et al., 2000). To evaluate problems associated with global climate change, floods, earthquakes, storms, landslides, and other environmental hazards, the behavior of the natural environment must be taken into account.

Understanding the intricate interactions amongst the natural environment and the built environment is crucial in encouraging the economic utilization of waste and energy reduction (Gu et al., 2006; Ding et al., 2009).

2.2. HISTORICAL AND PHILOSOPHICAL BACKGROUND

The Industrial Revolution manifested the turning point in the history of humankind. By the end of the 1700s, the manual labor-based economy was increasingly substituted by the industry and manufacture of machinery. Large scale production of the chemicals considerably contributed to economic development and the associated pollution of the natural environment brought about by the Industrial Revolution, with the creation of hydrochloric acid, sulfuric acid, and alkalis. Trade expansion was aided by the introduction of the railway systems. Steam-powered machinery produced dramatic increases in the production capacity and technological developments and the engineering breakthroughs of the Industrial Revolution encouraged urbanization. The quick growth in the urban populations augmented the demands for yield from usually outside the cities and the ecological impression of the whole countries rapidly reached beyond their boundaries, leading the path towards the modern global economy (Landes, 2003). Some philosophers responded to the Industrial Revolution via writings that emphasized and reintroduced the profound association between nature and man. The book of Ralph Waldo Emerson's "*Nature*" is an example (Emerson & Bosco, 1971; Emerson, 1987). The book is a convincing essay in which Ralph Waldo Emerson describes nature as the all-encompassing entity which is fundamentally known to us instead of being merely the component of the outside world. Emerson's philosophy developed, especially in the United States, an approach of looking at the natural environment and man which placed nature at the center of humanity. Others, like Henry David Thoreau, were impacted by Emerson's writings. Emerson took the philosophical viewpoint which marked the significant point in the history of the science and, specifically, of the contemporary and biological neurosciences by encouraging the concept that the mental world and the natural world, or mechanisms of the human brain, are assured to have the common biological origins (Wang, 2004; Dietz et al., 2009; Daw et al., 2011). Both Thoreau's and Emerson's writings contributed considerably to environmentalism as a political and a philosophy and social movement linked with the improvement and conservation of the natural environment. Later, the book of Aldo Leopold "*A Sand County*

Almanac" (Leopold, 2004) became the main reference that strengthened the ethical perspective of the environmentalist philosophy, disagreeing that it is immoral to damage the natural environment and that the humankind has an ethical responsibility to respect and safeguard it. The ethical viewpoint has significantly impacted the science of contemporary economics and environmental engineering (Gunter, 1990; Common, 2002). Whereas the engineering community exhibited little interest in environmental problems for a long time after Industrial Revolution, this condition altered with Richard Buckminster Fuller, an innovative engineer and scholar who was the first to encourage environmental issues as the main topic of both applied and fundamental science (Blignaut, 2004; Arrow et al., 2010).

2.3. ENGINEERING FOR LIVINGRY

Richard Buckminster Fuller born in Massachusetts in 1895 and, expelled from Harvard University as the student, consequently became a protruding early environmental scholar and activist, starting and cooperating in inventive design projects with the professionals, scientists, and artist worldwide. An engineer, designer, developer, architect, scholar, and scientist, he registered a large number of United States patents for his inventions. Buckminster Fuller lectured at the leading Universities around the entire globe and dedicated his teachings and professional practice to applying the principles of science and engineering to solve issues of society, aware that humankind would have to depend on the renewable sources of energy (Donofrio, 2018). His protective vision of planet running dry of the natural resources was established by MEA (*Millennium Ecosystems Assessment)* no later than 2 decades after his death, with the inference that the human activity over the decades has changed ecosystems more quickly and broadly and speeded up global climate change more prominently than in any similar time span in the entire history of man. Richard Buckminster Fuller encouraged the systemic method of environmental problems (Piroozfar & Farr, 2013). His work discovered solutions for material and energy-efficient design and engineering like foldable emergency shelters, lightweight structures of buildings, recycling aerodynamic automobiles and renewable resources, wind energy solutions for the large buildings and single households shower and tap systems that help decrease water consumption (Burger, 1965a; 1965b).

Fuller's prototype of the curve-shaped, low-cost, modular, energy-efficient and movable living structure designed in1940s, the Dymaxion House (Figure. 2.1), was equipped with advanced technology using natural winds for ventilation and cooling and featuring individual showers and taps with the economic spray systems in order to help decrease water consumption.

Figure 2.1: A prototype of Richard Buckminster Fuller of energy effective individual housing, the "Dymaxion House," is demonstrated at Henry Ford Museum in Dearborn, Michigan, United States.

[Source: https://www.pinterest.com/tunawidow85/tomorrowland/]

Buckminster Fuller and Kuromiya (1981) in their book "Critical path" passionately and resolutely advocates the concept of doing more with less. He debates for economic and efficient utilization of the natural resources and endorses the hunt for solutions that withstand the expansion of all forms of life on Earth. He can be considered as the father of engineering for sustainability and his philosophy was certainly the forerunner of the modern anthropocentric tactic (Aronson et al., 2006) to environmental sciences. His teachings and entire conception of education were dedicated to raising the growth of the life-long skills covering the extensive range of competence and knowledge to accomplish what he called the "*Omni*-successful education." The concept of livingry, which Fuller devised as opposed to weaponry, laid the theoretical basis on which the contemporary concept of sustainable development is based (Hoffert et al., 2002). His integrative view of the engineering science can normally be summarized by the schematic model displayed in Figure 2.2 below.

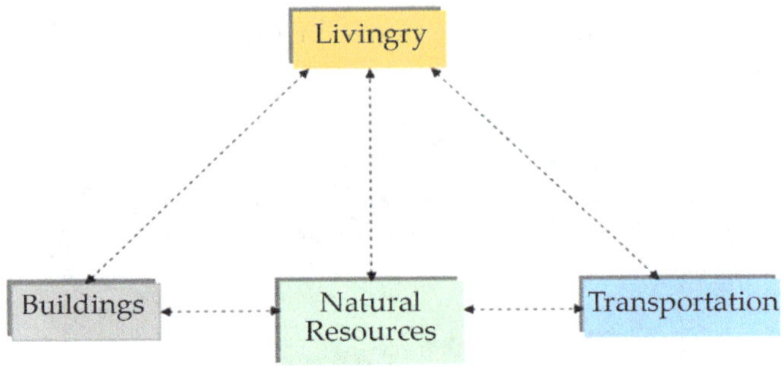

Figure 2.2: Fuller's world-view is considered as a first anthropocentric idea of the built and natural environment, where manmade transport and facilities exhaust and exploit raw materials.

[Source: https://benthamopen.com/contents/pdf/TOENVIEJ/TOENVIEJ-1–1.pdf]

Fuller's world-view can generally be considered as a first anthropocentric idea of the built and natural environment, where manmade transport and facilities exhaust and exploit raw materials and the natural energy resources like water and air, which are crucial for supporting human and the other life on Earth. Fuller anticipated that natural resources would be endangered by pollution and over-exploitation. He is regarded to be among the first engineering scientists to have actively and consistently encouraged waste recycling and renewable energy solutions.

The influence of the human-engineered systems on the natural environment has become the global problem of the highest significance since the death of Fuller in 1983. The grander objectives of modern engineering have been concerned with remediation and anticipation of environmental pollution and exhaustion of the raw materials and some other natural resources (Rosen, 1983).

2.4. A JOURNEY FROM STUDIES OF THE ECOSYSTEMS TO THE ENVIRONMENTAL ENGINEERING

Systems ecology bloomed from the 1960s as the discipline within biological sciences. Ecosystem approaches donated to upsurge understanding of how the atmosphere reacts or responses to the joined effects of geochemical recycling and global metabolism, and how the biogeochemical procedures are in turn impacted by variations in the atmosphere (Gaudy & Gaudy, 1980; Khan & Ansari, 2005). The seminal books by Eugene Odum (1959) and Howard Odum (1971) offered a complete analysis of the fundamentals and history of the ecology of the system. Systemic methods in the ecology draw from an extensive range of the scientific disciplines comprising geology, conservation biology, biochemistry, and evolutionary ecology. Some authors (Callicott et al., 1999) have tried to further divide the field of ecology into methods concerned with conservation and biodiversity of the ecological entities, and more systemic methods marking at understanding procedures of the ecosystem functioning. The reality of existing research in the field, though, displays that integrative forms of ecological research are evolving, blurring the lines amongst specialized domains beyond and within ecology. Theoretical models incline towards progressively global biosphere methods, marking at the understanding of Earth as a whole. In a model of the balanced global biosphere, photosynthesis fixes carbon dioxide (CO_2) and regenerates oxygen whereas humans, plants, animals, and microbes breathe from the atmosphere by utilizing oxygen and generating carbon dioxide at the same rate (Odum & Odum, 2003). In the world, such kind of balanced behavior no longer exists. The main challenge for the ecology now is to acquire an understanding of the large-scale effects leading to imbalanced behavior in the global biosphere and to further the knowledge regarding underlying biogeochemical procedures and how they interact. Engineering science has attained a critically vital role in this attempt (Graham & Smith, 2004; Erwin, 2008). Solutions to the pressing issues of an imbalanced Earth are anticipated to outcome from sufficient engineering development and technology, centered on the understanding of observational data and research models (Wright & Jones, 2006). Theoreticians have previously asserted on differences amongst environmental engineering and ecological engineering (Allen et al., 2003), the latter being viewed as fundamentally concerned with restrictions on living organisms and ecosystems, whereas the former would chiefly address constraints on the non-living environments and would thus be more closely recognized with conventional engineering (Gattie et al., 2003; Wright et al., 2006). The concept of the theoretical separation amongst topics and issues of ecological engineering and environmental engineering is now problematic to reconcile with modern engineering, in terms of both research and practice. The need for substituting the complete domain of science within the context of urgency has developed a situation where conventional engineering is the thing of the past. With the outline of the concept of sustainability in terms of the global challenge (Painter, 2003) hitting any field of modern science and technology and thus reaching well above classic ecology, a substantial change in thinking has taken place. Global anthropocentric theories have started to tighten prevailing links between the scientific domains. These include links amongst mathematics, economics and computer modeling, and links amongst the social sciences and the humanities to the advantage of environmental science (Painter, 2003).

2.5. SUSTAINABLE GLOBAL CHANGE

Whereas the practice of environmental engineering has progressed significantly during the past 2 decades, its fundamental contract with society has been the same. Environmental engineering is about offering a quality of life, with the safeguard, nurture and restitution of the fragile environment in mind. Environmental scientists forecast that the individual houses and local communities will usually become smaller and will be centered on principles of the ecological design (Farley & Costanza, 2002; Todd et al., 2003), where refurbishment of the natural environments within the urban landscapes is to perform an important role. The ecological method to the built environment comprises a holistic design method where energy utilization and resource depletion is decreased, and internal and external pollution causing potential hazards to the users and the environment is minimized. An instance of such kind design at a level of the individual housing is the Hockerton project, which was finished in 1996 in Great Britain and registered in the list of best practice examples by the European Green Building Forum in 2001 (Ghosh & Vale, 2006). In this type of project, the rainwater from the house roofs is filtered and gathered as drinking water. As plastic is probable to produce toxins, the copper gutters are utilized. To save energy, the water is usually heated with a pump. Water falling somewhere else on the estate is then pumped to the reservoir which is large enough to hold supply for several months for growing crops and washing laundry. The capacity of the reservoir is around 150 meters and the water runs via a natural filter system. All of the sewage treatment systems are usually biological and rely completely on natural organisms (De Haan, 2000; Voisey & O'Riordan, 2010). The mechanical aeration system generally recycles air and conserves heat within the building. The cost of heating is minimized. Figure 2.3 displays how the houses in the Hockerton project were sheltered with earth to preserve heat. The sustainable design of a building is the thoughtful incorporation of ecologically conscious architecture and mechanical, electrical, and structural engineering. In addition to the concerns for proportion, scale, aesthetics, light and texture, the costs of an ecological long term are considered. Connecting design of building with nature fetches new forms of life to the urban dwellings and reminds of the place within nature. Sustainable design of a building is usually to take into concern an extensive range of races, religions, cultures and habits in order to meet the requirements of communities and individuals, at global and local scales (Escobar, 2011; Herrero & Thornton, 2013).

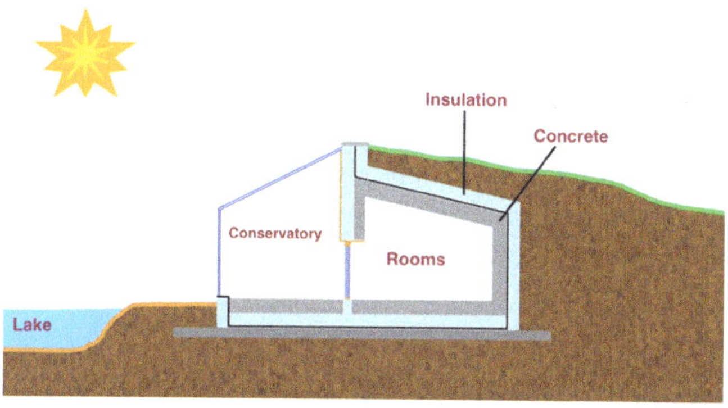

Figure 2.3: In Hockerton Project roofs were generally covered with soil instead of tiles to upsurge energy efficiency by trapping the heat. The soil plays the role of a giant radiator.

[Source: https://ebuilding.blog/hockerton-housing-project]

The subject of environmental engineering has to face the test of exceptional global industrialization and urbanization in developing countries like Africa, which pursue to increase their living standards and wealth. Engineers nowadays must be able to acclimate their vision and react to the necessities of societies at a more complex and much larger scale, where collaboration within and with other cultures is needed (Hall, 2010; DeFries et al., 2012). Among the most crucial demands presently put to the profession, the following may be listed:

1. Educate the public in order to increase consciousness of the environmental issues;
2. Mitigate the dangers of global climate change;

3. Provide sufficient supplies of the freshwater;
4. To deal with pressures of the increasing population;
5. Expect energy shortages, searching for renewable energy sources;
6. Develop tools to evaluate the performance;
7. Sustainability of the existing structures;
8. Design ecologically welcoming transportation systems estimate influences and expect risks;
9. Restore and reclaim disturbed and mined landscapes, develop and strengthen associations with the other fields of science like mechanics, the geological, biochemical engineering, biological, social, computer and management sciences.

These complex problems cut across all the domains of engineering and need a scientific analysis at the different levels of complexity and scale. Management training and proficiency to put novel solutions into practice is a necessary aspect of modern environmental engineering. (Figure 2.4) suggests the schematic view of issues to be addressed, at large and small scales (Redclift, 1992).

2.6. ADAPTIVE PROJECT MANAGEMENT FOR COMPLEXITY AND SCALE PROBLEMS

A factor that is frequently difficult to evaluate in search of sustainable solutions is the complexity and scale of the given problem. Engineers might be perfectly proficient in working out the simple physical design or hydro-geomorphology for what might seem a suitable solution to the problem of environment, but might eventually not provide an effective solution because the complexity or scale of the problem wasn't sufficiently predicted and analyzed. The scale and complexity of the biological systems (Engel & Odum, 1999) make it very difficult, often impossible, to comprehend how the systems act at different levels of the organization from elemental cycles to the community metabolisms and how these levels cooperate. Furthermore, biological systems and the components take variable and different times to react to the engineering solutions or to variations in general (Jax, 2006). The timescale of the responses might exceed that of the reasonable monitoring anticipations and might render consistent predictions impossible. This timescale issue can be demonstrated by the example of impaired coastal reefs, where the times for full functional refurbishment at the site might be assessed in terms of the centuries. Also, biological systems react to the variations as they develop and this natural adaptation might make it impossible or difficult to work out the ready regulatory recipes. A different but associated issue of scale in environmental engineering is that linked to a number of people affected or concerned by the project. Adaptive project management was suggested as a method that helps cope with the problems of scale (Hackney, 2000; Thom, 2000).

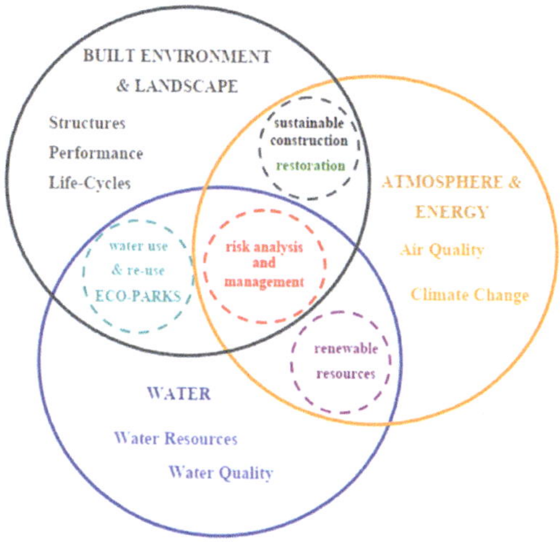

Figure 2.4: The built environment reveals the material wealth of nations and cities. Its sustainability depends largely on the rigor and care with which the resources are utilized and reused.

[Source: https://benthamopen.com/contents/pdf/TOENVIEJ/TOENVIEJ-1-1.pdf]

Understanding interactions amongst the built environment and the natural context is necessary for encouraging an economic utilization of resources. Finding sustainable solutions comprises research for the renewable resources and analysis and management of the risks like global climate change, floods, earthquakes or storms. The accessibility and consumption of the energy impact both the extent and nature of the pollution and the amount to which the polluted atmosphere or contaminated sites can be restored or regenerated. Countries with poorly established infrastructures in which the supply of energy is rare will have to deal with the environmental problems at a different scale compared with the countries which have highly established infrastructures and broad energy production and usage.

Adaptive management permits for changing objectives as an engineering project progresses with time. Intelligent monitoring joined with the general project goals permit accurate re-evaluations of what can be achieved. Adaptive management works finest with the conceptual model that inclines to clearly defined, unambiguous general goals (Marcus, 2000). The victory criteria in the adaptive project management change from case to case and the more victory criteria are dependent on function and process *per se*. For procedures that are well implicit, a list of comprehensive victory criteria can be developed from the outset. Most significantly, adaptive management allows the learning process intrinsic to contemporary engineering research and practice to be gradually integrated into the given project. As far as the planning is considered, any well-established project plan must include the flow chart with branches at the points in a procedure where uncertainty prevails and where eccentricity from an anticipated result path leads to an expected alternative endpoint. Well-designed engineering projects integrate uncertainty into the procedure and at the outset enlighten the responsible party or public of the potential problems, substitute solutions, and the cost.

These are signified by alternative endpoints and pathways in a project flow chart. Adaptive management permits all parties involved in order to consider the range of potential end-products and decide if these products are acceptable.

2.7. ASSESSMENT OF DAMAGE AND HARM OF RESOURCES

Landscape planning and policy-making are exaggerated by the influence of both natural and man-made caused harm to the environment. As an instance, the case of the earthquakes may be considered, such as that which took place in September 1999 in Central Taiwan, inducing various large-scale landslides and triggering resource and environmental problems that needed the functional analytical ecosystem method to the issue outside the standard technological solutions. The influence study by Lin et al. (2006) gives an instance of the long-term examination into the concerns of the earthquakes on the environment. In this study, variations of the post-quake landslides, soil loss, and vegetation recovery conditions were assessed utilizing imaging technology like multi-temporal SPOT satellite images and the analytical image transformation. Mathematical models for the vegetation recovery analysis centered on calculations of NDVI (normalized difference vegetation index) and the Universal Soil Loss Equations (Wilson, 1989) directed the research and evaluation. Landslides were acknowledged on the basis of post- and pre-quake air images utilizing image differencing algorithms and variations detection thresholds. VRR (Vegetation recovery rates) were calculated to compute successive conditions of the vegetation recovery at the landslides. To approximate the environmental influence of denudation sites, the loss of soil before and after the earthquake was computed on the basis of the geomorphology analysis. The research exposed the large-scale influence of earthquake and the concomitant landslides and permitted further assessment making policies and planning in affected areas. After 6 years of monitoring, the vegetation recovery was discovered to have reached 89.69 percent, which provides an overall assessment of nature's ability to renew vegetation after the large-scale landslides. However, other factors of risk like typhoons could eliminate such renewal in a short time. To diminish such kinds of risks, engineering solutions for the geo-morphologically unstable sites must be developed. Other large and medium-scale risks that challenge the proficiency of practitioners and researchers in the field of environmental engineering are the ones signified by the pollution of water supplies via toxic trace elements.

2.8. RISK EVALUATION AT VARIOUS ORGANIZATIONAL LEVELS

Insufficient water supply and associated health dangers are the daily issues faced by millions of people worldwide. The management of the water environment for sustainable human benefit needs environmental policies that encourage ecological health, meaning both human safety and ecosystem health. The worries of environmental engineers for the water supplies, the vindication of pollution, and the encouragement of the ecological health nowadays spread over the safety and performance of the individual components. Whole water supply systems and collaborations with the other systems are considered, which comprises assessing the risks and examining the results of water pollution by toxic trace elements or pathogens. These might originate from several sources. Metal traces in the soils, for instance, were discovered to cause potentially pollutant effects at all levels of the biological organization, from cellular to the ecosystem levels, even in the sites where the conforming surface water met the criteria of water quality (Peplow & Edmonds, 2005). The existence of COPECs or pollutants of the potential environmental concern in water and soil signifies an ecological risk at the ecosystem, population, individual, community, cellular, and molecular levels (Calow, 1998). Whereas the detection and recognition of COPECs are, in principle, not very difficult, the current issue with toxicity risk assessment is that the specificity of pollutant effects and the insight expanded into the mechanisms of the toxicity as such is quite lesser the higher level of the organization (Hodson, 1990). Indicators of the toxicity like morphological changes at the tissue level, ultra-structural variations at cellular level exposed via electron microscopy and the biochemical changes at the molecular level allow establishing and to computing cause-effect relationships. Evaluation of the contamination risks at the level of ecologically appropriate procedures is far more problematic and vulnerable to biases because of the lack of the individual data on outcomes, exposure, or confounding the variables that might contribute to the measured effect (Clements, 2000). At the lower levels of an organization, crucial changes take place more quickly and might offer early warnings of the toxicological effects on the populations (Tamaki & Frankenberger, 1992). However, in spite of the greater mechanistic consideration and specificity of the effects at the lower levels of an organization, the understanding offered may be limited as the importance of the effects at an ecological level or the largeness of the bio-magnification is unidentified. The practicality of indicators, therefore, depends on the cross-examination of the various levels of an organization. The effects of the trace elements on the plant communities, for instance, might be relevant as the production of the plant matter is the main source of the organic carbon for the aquatic ecosystems. A probable food-chain transfer from polluted soils must also be considered. Changes at the ecosystem level might be inferred when pollutants exceed the benchmark levels that are poisonous to soil bacteria, signifying that the functional properties linked to nutrient cycling and the energy flow have normally been affected (Dallinger & Kautzky, 1985; Hare, 1992). The paths for migration of the pollutants away from the given source might involve transport in the shallow groundwater and surface water (Dresp, 2008). Although dissolved the metal trace elements in the water might appear mild with respect to poisonousness, this doesn't lessen anxieties over oral uptake of the trace elements via contaminated food. It is normally believed that the acceptance of the adsorbed trace elements is considerably less hazardous than the absorption of the dissolved forms. However, the comparative significance of different routes of the exposure stays unclear and at the high concentrations bioavailability of even the small fraction of the adsorbed trace elements from the diet might be significant, which would assist the hypothesis that the diet is a substantial route of exposure (Dallinger & Kautzky, 1985; Hare, 1992). At the level of the human population, the danger of death from cancer is projected to be augmented in people reveled to trace elements via diet, even at the seemingly low concentrations. For the last thirty years, eco-toxicology has been related to studying the impacts of pollutants on the ecological health and biological integrity of the communities and ecosystems. Newman and (2014) gave a complete and brief overview, from general principles of the eco-toxicology to the examples of scientific in-depth studies. Some of the examples are concerned with the effects of the environmental contaminants at different levels of the biological organization, from molecule to biosphere as a whole. How *biotic veracity* can be evaluated at different levels and utilized as the monitoring tool in order to guide the additional investigation, is discussed. To reinstate the ecological health of soils, water, and sediments, environmental and ecological engineering design solutions are broadly applied to the gradually diverse range of groundbreaking projects for efficient management of the natural resources. Sufficient technologies have been developed and are presently utilized within the fields of environmental restoration and protection, industrial waste and sewage treatment, the food sector, architecture, and landscape design.

2.9. TURN-KEY APPROACHES FOR RESTORATION OF THE ECOLOGICAL HEALTH

Environmental engineering practice is about what Aronson *et al* (2006) called the quest for efficient turnkey approaches to diminish or invert the negative effects of the environment of the Twentieth century's growth path. A turnkey approach, in business language, is a solution or strategy where the designer or provider undertakes the complete charge from design to the completion and commissioning. Several concrete examples of the environmental turn-key solutions at a medium scale are provided in the circumstance examples given by Todd, Brown, & Wells (2003).

The economic feasibility of integrative ecosystems and the environmental engineering design within the urban settings is a subject that distresses both the public and the local authorities involved in decision making and planning (Aronson et al., 2006b). Urban environments offer the foundation for most of the economic activities of the world, including leisure and the tourism industry. Urban planning and development usually have a major influence on the economic performance of the companies, nations, and corporations, affecting human activities (Heal, 2000). Demographic approximates of a large number of the surplus residents of the planet for the next decade need planning and development of the basic infrastructures and housing, which is to suffer a gigantic cost in order to support the anticipated population growth. The infrastructures are quickly decaying in several countries and are immediately in the necessity of repair and maintenance, which signifies projects linked with another massive outlay. The escalating trans-national embedding of the global sustainability problems will lead to the private-public corporations and the projects that will be gentle to manage, particularly in the developing countries that don't have the convention of private ownership and the public infrastructure. Therefore, both professionals and scientists in the field of environmental engineering will generally be confronted with extraordinary social and economic challenges.

2.10. ENVIRONMENTAL ENGINEERING FOR THE SUSTAINABLE ECONOMIC DEVELOPMENT

Whether an inventive design solution that welfares the environment and people will eventually meet success dependent on economic and social factors. Questions of the cost-effectiveness or whether the people are probable to admit and live with the solution offered to them must be addressed. Idealistic apparitions have to lead to genuine results and ethical anxieties are considered in the light of political and economic pressure. This comprises cost-benefit analyses, viability studies and consciousness of political climate and the decision-making framework as potential controlling factors to the project. The emergent discipline of the ecological economics discourses such intricate issues at the interface of economic and the global ecosystem that withstands and contains it. Like environmental engineering, ecological economics adopts a systems tactic that goals to provide solutions to issues of the society in the 21st century (Mihelcic et al., 2006; Yaun et al., 2006). Economics is usually defined as a science of the allocation of rare resources among substitute desirable ends. This implies a precise sequence of the steps in the economic examination and, lastly, the economist should decide what ends must be pursued. Many argue that conclusions on ends must be left to the democratic processes, proposing that a necessary precursor to the economic analysis is a democratic process that effectively articulates the anticipated goals. Once these later have been acknowledged, the economist can examine what resources are essential to accomplish them, and which are the scarcest. The concluding step is allocation, through whatever mechanism or institution is most suitable for resources and the ends considered (Allenby et al., 2009; Haapala et al., 2013). In classical economics, there isn't a complete resource scarcity. As the resource becomes rare, its cost increases, which offers market incentives to grow substitutes. The issue with such thought is that the humans and other species depend on for their welfare and survival on the hitherto non-marketed ecosystem services like climate, water, air, and the clean atmosphere. Though increasingly rare, most of the services have no expense and therefore don't produce any feedback from the markets to specify their scarcity. Theoretical economics focuses on markets and on growths, but markets flop to efficiently assign most natural resources, probably with the exclusion of water (F. Lozano & R. Lozano, 2014). The critical issue is that, as the economy grows, it doesn't expand into the void, but into the finite sustaining and comprising biosphere. Throughout the procedure of economic growth, society uses natural capital, and thus its crucial support functions. The possible irreversibility of this procedure has often been debated on the instance of the energy consumption and the second law of thermodynamics (Georgescu-Roegen, 1971). The universal law of growing entropy or increasing inaccessibility of energy in the physical system forecasts that in any remote system entropy increases. Applied to the economic process, as discoursed by Georgescu-Roegen (1971), the second law of the thermodynamics leads to forecast that, eventually, the rarest resource of energy will be the low-entropy matter-energy. The aptitude of societies to depend on such kind of energy would, therefore, eventually, be limited. New theories of the economic development, which integrate concepts of the development ethic and dematerialization, have been offered, like the Robert Ayres' theory, which asserts on the difference amongst more and better, and which trials the classic development example by pointing out that the appropriate measure of the economic output isn't the number of goods manufactured, but the quality and, not essentially material, value of the final services given to the consumer (Farley & Daly, 2006).

Modern environmental engineering is regarding developing systems, methods, structures, tools, and infrastructures that aid to protect the environment and human health, comprising sustainable transportation and the energy systems and solutions for adequate and safe water supply worldwide. Problems need to be examined on a large scale, considering the rising trans-national embedding of environmental research and technology. The examples discoursed here demonstrate how the context, research, and practice of environmental engineering, in close contact with the other disciplines like ecology, geology, economics, and biology, address the urgent issues presently faced by the global society.

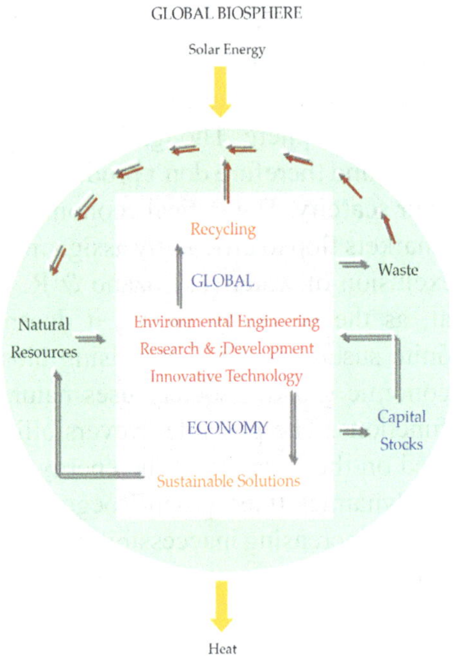

Figure 2.5: A growing economy doesn't expand into the void, however, into the finite sustaining and comprising biosphere. Through the procedure of economic growth, society uses natural capital and thus its energetic support functions.

[Source: https://benthamopen.com/contents/pdf/TOENVIEJ/TOENVIEJ-1-1.pdf]

Contemporary economic science searches for the models of economic development skilled in assigning resources in a sustainable and efficient way, where human welfare is continued, natural resources are endangered and reinstated, and waste production is decreased. This should comprise the sound but enthusiastic re-investment of the capital stocks produced via economic development into research and groundbreaking technology encouraging new, sustainable solutions

Understanding complex systems need more than one model and there is a rising agreement on the necessity for altercation across the engineering, natural, and social sciences in order to address questions of the economic development and sustainability, the increasing shortage of the natural resources and the future of biosphere as a whole (Ayres, 1998). The necessity for the new engineering solutions is crucial, the stakes are very high. By considering the psychological and social implications of the priorities and by encouraging exchanges amongst scientists across the disciplines with the public, it is hoped to be able to upsurge to the challenge (Brown & Ulgiati, 1997; Mihelcic et al., 2008).

REFERENCES

1. Allen, T. F., Giampietro, M., & Little, A. M. (2003). Distinguishing ecological engineering from environmental engineering. *Ecological Engineering*, *20*(5), 389–407.
2. Allenby, B., Murphy, C. F., Allen, D., & Davidson, C. (2009). Sustainable engineering education in the United States. *Sustainability Science*, *4*(1), 7.
3. Aronson, J., Blignaut, J. N., Milton, S. J., & Clewell, A. F. (2006a). Natural Capital: The Limiting Factor, 1, 1–30.
4. Aronson, J., Milton, S. J., Blignaut, J. N., & Clewell, A. F. (2006b). Nature conservation as if people mattered. Journal for Nature Conservation, 14(3–4), 260–263.
5. Arrow, K. J., Dasgupta, P., Goulder, L. H., Mumford, K., & Oleson, K. (2010). China, the US, and sustainability: Perspectives based on comprehensive wealth. In *Is Economic Growth Sustainable?* (Vol. 1, pp. 92–144). Palgrave Macmillan, London.
6. Ayres, R. U. (1998). Turning Point: The End of the Growth Paradigm (Vol. 11, pp. 10–60). *First published*.
7. Bandura, A. (2001). Social cognitive theory: An agentic perspective. *Annual Review of Psychology*, *52*(1), 1–26.
8. Blignaut, J. N. (2004). Towards an economic development ethic. *Sustainable Options*, 1, 515–522.
9. Brown, M. T., & Ulgiati, S. (1997). Emergy-based indices and ratios to evaluate sustainability: monitoring economies and technology toward environmentally sound innovation. *Ecological Engineering*, *9*(1–2), 51–69.
10. Burger, D. H. (1965a). World Design Science Decade, 1965–1975: Phase I (1963) Document 1, Inventory of World Resources, Human Trends, and Needs by R. Buckminster Fuller and John McHale (Book Review). *Technology and Culture*, *6*(3), 504.
11. Burger, D. H. (1965b). World Design Science Decade, 1965–1975: Phase I (1964) Document 2, The Design Initiative by R. Buckminster Fuller and John McHale (Book Review). *Technology and Culture*, *6*(3), 504.
12. Callicott, J. B., Crowder, L. B., & Mumford, K. (1999). Current normative concepts in conservation. *Conservation Biology*, *13*(1), 22–35.

13. Calow, P. (1998). Environmental risk assessment and management: the whats, whys and hows. *Handbook of Environmental Risk Assessment and Management,* P. Calow (ed.). Blackwell Science, Oxford, UK, Vol. 1(6), pp. 1–40.
14. Clements, W. H. (2000). Integrating effects of contaminants across levels of biological organization: an overview. *Journal of Aquatic Ecosystem Stress and Recovery*, *7*(2), 113–116.
15. Common, M. (2002). Partha Dasgupta, Human Well-Being and the Natural Environment. Oxford: Oxford University Press, 2001. ISBN 0–19–924788–9. xxii+ 305. *Journal of Public Policy*, *22*(3), 349–355.
16. Dalby, S. (2013). Climate change: New dimensions of environmental security. *The RUSI Journal*, *158*(3), 34–43.
17. Dallinger, R., & Kautzky, H. (1985). The importance of contaminated food for the uptake of heavy metals by rainbow trout (Salmo gairdneri): a field study. *Oecologia*, 1, 82–89.
18. Daly, H. E., & Farley, J. (2011). *Ecological Economics: Principles and Applications* (Vol. 1, pp. 1–50). Island press.
19. Daw, T. I. M., Brown, K., Rosendo, S., & Pomeroy, R. (2011). Applying the ecosystem services concept to poverty alleviation: the need to disaggregate human well-being. *Environmental Conservation*, *38*(4), 370–379.
20. De Haan, L. J. (2000). Globalization, localization and sustainable livelihood. *Sociologia Ruralis*, *40*(3), 339–365.
21. DeFries, R. S., Ellis, E. C., Chapin III, F. S., Matson, P. A., Turner, B. L., Agrawal, A., ... & Lambin, E. (2012). Planetary opportunities: a social contract for global change science to contribute to a sustainable future. *BioScience*, *62*(6), 603–606.
22. Dietz, T., Rosa, E. A., & York, R. (2009). Environmentally efficient well-being: rethinking sustainability as the relationship between human well-being and environmental impacts. *Human Ecology Review*, *1*, 114–123.
23. Ding, J. H., Xiao, K. Y., Lou, D. B., & Li, N. (2009). Large Scale Mineral Assessment in Three Dimensions [J]. *Geology and Exploration*, *6*, *1–20*.
24. Donofrio, M. K. (2018, July). A method of accounting for value at the former NAS Tillamook. In *Proceedings of IASS Annual Symposia* (Vol.

2018, No. 12, pp. 1–7). International Association for Shell and Spatial Structures (IASS).

25. Dresp, B. (2008). Dimensions of environmental engineering. *Open Environmental Engineering Journal, 1*, 1–8.
26. El-Fadel, M., Findikakis, A. N., & Leckie, J. O. (1997). Environmental impacts of solid waste landfilling. *Journal of Environmental Management, 50*(1), 1–25.
27. Emerson, R. W. (1987). *Representative Men: Seven Lectures* (Vol. 4, pp. 1–50). Harvard University Press.
28. Emerson, R. W., & Bosco, R. A. (1971). *The Collected Works of Ralph Waldo Emerson: Society and Solitude* (Vol. 7, pp. 10–60). Harvard University Press.
29. Engel, V. C., & Odum, H. T. (1999). Simulation of community metabolism and atmospheric carbon dioxide and oxygen concentrations in Biosphere 2. *Ecological Engineering, 13*(1–4), 107–134.
30. Erwin, D. H. (2008). Macroevolution of ecosystem engineering, niche construction and diversity. *Trends in Ecology & Evolution, 23*(6), 304–310.
31. Escobar, A. (2011). Sustainability: Design for the pluriverse. *Development, 54*(2), 137–140.
32. Farley, J., & Costanza, R. (2002). Envisioning shared goals for humanity: a detailed, shared vision of a sustainable and desirable USA in 2100. *Ecological Economics, 43*(2–3), 245–259.
33. Farley, J., & Daly, H. (2006). Natural capital: The limiting factor: A reply to Aronson, Blignaut, Milton and Clewell. *Ecological Engineering, 28*(1), 6–10.
34. Fuller, R. B., & Kuromiya, K. (1981). *Critical Path* (Vol. 1, pp. 1–30). Macmillan.
35. Gattie, D. K., Smith, M. C., Tollner, E. W., & McCutcheon, S. C. (2003). The emergence of ecological engineering as a discipline. *Ecological Engineering, 20*(5), 409–420.
36. Gaudy, A. F., & Gaudy, E. T. (1980). *Microbiology for Environmental Scientists and Engineers* (Vol. 736). New York: McGraw-Hill.
37. Georgescu-Roegen, N. (1971). The Entropy Lawandthe Economic Process Cambridge 1971. *Georgescu-RoegenThe Entropy Law and the Economic Process1971, 1, 10–40.*

38. Ghosh, S., & Vale, R. (2006). The potential for solar energy use in a New Zealand residential neighbourhood: a case study considering the effect on CO2 emissions and the possible benefits of changing roof form. *Australasian Journal of Environmental Management, 13*(4), 216–225.

39. Graham, D. W., & Smith, V. H. (2004). Designed ecosystem services: application of ecological principles in wastewater treatment engineering. *Frontiers in Ecology and the Environment, 2*(4), 199–206.

40. Gu, D., Zhu, Y., & Gu, L. (2006). Life cycle assessment for China building environment impacts. *Journal-Tsinghua University, 46*(12), 1953.

41. Gunter, P. A. (1990). For the Common Good: Redirecting the Economy Towards Community, the Environment, and a Sustainable Future. *Process Studies, 19*(1), 56–61.

42. Haapala, K. R., Zhao, F., Camelio, J., Sutherland, J. W., Skerlos, S. J., Dornfeld, D. A., ... & Rickli, J. L. (2013). A review of engineering research in sustainable manufacturing. *Journal of Manufacturing Science and Engineering, 135*(4), 041013.

43. Hackney, C. T. (2000). Restoration of coastal habitats: expectation and reality. *Ecological Engineering, 15*(3–4), 165–170.

44. Hall, C. M. (2010). Changing paradigms and global change: From sustainable to steady-state tourism. *Tourism Recreation Research, 35*(2), 131–143.

45. Hare, L. (1992). Aquatic insects and trace metals: bioavailability, bioaccumulation, and toxicity. *Critical Reviews in Toxicology, 22*(5–6), 327–369.

46. Heal, G. (2000). *Nature and the Marketplace: Capturing the Value of Ecosystem Services* (Vol. 1, pp. 11–30. Island Press.

47. Herrero, M., & Thornton, P. K. (2013). Livestock and global change: emerging issues for sustainable food systems. *Proceedings of the National Academy of Sciences, 110*(52), 20878–20881.

48. Hodson, P. V. (1990). Indicators of ecosystem health at the species level and the example of selenium effects on fish. *Environmental monitoring and Assessment, 15*(3), 241–254.

49. Hoffert, M. I., Caldeira, K., Benford, G., Criswell, D. R., Green, C., Herzog, H., ... & Lightfoot, H. D. (2002). Advanced technology paths

to global climate stability: energy for a greenhouse planet. *Science, 298*(5595), 981–987.

50. Jabbour, C. J. C., & Santos, F. C. A. (2008). Relationships between human resource dimensions and environmental management in companies: proposal of a model. *Journal of Cleaner Production, 16*(1), 51–58.
51. Jax, K. (2006). Ecological units: definitions and application. *The Quarterly Review of Biology, 81*(3), 237–258.
52. Khan, F. A., & Ansari, A. A. (2005). Eutrophication: an ecological vision. *The Botanical Review, 71*(4), 449–482.
53. Landes, D. S. (2003). *The Unbound Prometheus: Technological Change and Industrial Development in Western Europe from 1750 to the Present* (Vol. 1, pp. 10–40). Cambridge University Press.
54. Leopold, A. (2004). A Sand County Almanac. *This Sacred Earth: Religion, Nature, Environment, 179, 36.*
55. Lin, W. T., Lin, C. Y., & Chou, W. C. (2006). Assessment of vegetation recovery and soil erosion at landslides caused by a catastrophic earthquake: a case study in Central Taiwan. *Ecological Engineering, 28*(1), 79–89.
56. Lozano, F. J., & Lozano, R. (2014). Developing the curriculum for a new Bachelor's degree in Engineering for Sustainable Development. *Journal of Cleaner Production, 64*, 136–146.
57. Marcus, L. (2000). Restoring tidal wetlands at Sonoma Baylands, San Francisco Bay, California. *Ecological Engineering, 15*(3–4), 373–383.
58. Mihelcic, J. R., Paterson, K. G., Phillips, L. D., Zhang, Q., Watkins, D. W., Barkdoll, B. D., … & Hokanson, D. R. (2008). Educating engineers in the sustainable futures model with a global perspective. *Civil Engineering and Environmental Systems, 25*(4), 255–263.
59. Mihelcic, J. R., Phillips, L. D., & Watkins Jr, D. W. (2006). Integrating a global perspective into education and research: Engineering international sustainable development. *Environmental Engineering Science, 23*(3), 426–438.
60. Newman, M. C. (2014). *Fundamentals of Ecotoxicology: The Science of Pollution* (Vol. 1, pp. 7–100). CRC Press.
61. Noguera, D. R., Pizarfo, G., Stahl, D. A., & Rittmann, B. E. (1999). Simulation of multispecies biofilm development in three dimensions. *Water Science and Technology, 39*(7), 123–130.

62. Odum, E. P. (1959). Fundamentals of ecology: Vol. xvii, pp. 1–546, frontispiece.
63. Odum, H. T. (1971). *Environment, Power, and Society* (Vol. 1, pp. 1–50). New York, NY: Wiley-Interscience.
64. Odum, H. T., & Odum, B. (2003). Concepts and methods of ecological engineering. *Ecological Engineering*, *20*(5), 339–361.
65. Painter, D. J. (2003). Forty-nine shades of green: ecology and sustainability in the academic formation of engineers. *Ecological Engineering*, *20*(4), 267–273.
66. Park, K., Hwang, Y., Seo, S., & Seo, H. (2003). Quantitative assessment of environmental impacts on life cycle of highways. *Journal of Construction Engineering and Management*, *129*(1), 25–31.
67. Peplow, D., & Edmonds, R. (2005). The effects of mine waste contamination at multiple levels of biological organization. *Ecological Engineering*, *24*(1–2), 101–119.
68. Piroozfar, P., & Farr, E. R. (2013). Evolution of nontraditional methods of construction: 21st century pragmatic viewpoint. *Journal of Architectural Engineering*, *19*(2), 119–133.
69. Redclift, M. (1992). Sustainable development and global environmental change: implications of a changing agenda. *Global Environmental Change*, *2*(1), 32–42.
70. Rosen, S. (1983). The role of the history of science in the training of science teachers., etc.*: A Review of General Semantics*, 1, 471–477.
71. Salic, A., & Zelic, B. (2018). Introduction to environmental engineering. *Physical Sciences Reviews*, *3, 1–22.*
72. Tamaki, S., & Frankenberger, W. T. (1992). Environmental biochemistry of arsenic. In *Reviews of Environmental Contamination and Toxicology* (Vol. 1, pp. 79–110). Springer, New York, NY.
73. Thom, R. M. (2000). Adaptive management of coastal ecosystem restoration projects. *Ecological Engineering*, *15*(3–4), 365–372.
74. Todd, J., Brown, E. J., & Wells, E. (2003). Ecological design applied. *Ecological Engineering*, *20*(5), 421–440.
75. Voisey, H., & O'Riordan, T. (2010). Globalization and localization. In *Globalism, Localism and Identity* (Vol. 1, pp. 44–61). Routledge.
76. Wang, S. (2004). One hundred faces of sustainable forest management. *Forest Policy and Economics*, *6*(3–4), 205–213.

77. Wilson, J. P. (1989). Soil Erosion From Agricultural Land in the Lake Semcoe–Couchiching Basin, 1800–1981. *Canadian Journal of Soil Science*, *69*(1), 137–151.
78. Wright, J. P., & Jones, C. G. (2006). The concept of organisms as ecosystem engineers ten years on: progress, limitations, and challenges. *BioScience*, *56*(3), 203–209.
79. Wright, J. P., Jones, C. G., Boeken, B., & Shachak, M. (2006). Predictability of ecosystem engineering effects on species richness across environmental variability and spatial scales. *Journal of Ecology*, *94*(4), 815–824.
80. Young, D., Scharp, R., & Cabezas, H. (2000). The waste reduction (WAR) algorithm: environmental impacts, energy consumption, and engineering economics. *Waste Management*, *20*(8), 605–615.
81. Yuan, Z., Bi, J., & Moriguichi, Y. (2006). The circular economy: A new development strategy in China. *Journal of Industrial Ecology*, *10*(1–2), 4–8.s

CHAPTER 3

Green Manufacturing Technologies

CONTENTS

3.1. INTRODUCTION

This chapter gives an overview of green manufacturing technologies. The details about different types of eco-friendly and modern manufacturing expertise that are being adept by several industries are discussed. The importance and steps which need to be followed for implementing green manufacturing expertise are synopsized. The necessity and use of environmentally friendly manufacturing expertise are primarily focused upon (Velayudham, 2007; Karana et al., 2009). The need for substituting prevailing manufacturing technology for specific industries with eco-friendly techniques is emphasized. The models of different companies around the globe executing green manufacturing technologies are also discussed. The savings of energy ingestion by utilizing renewable energy resources is then compared to that of the conventional energy resources (Murugesan, 2008; Kumar & Rao, 2018).

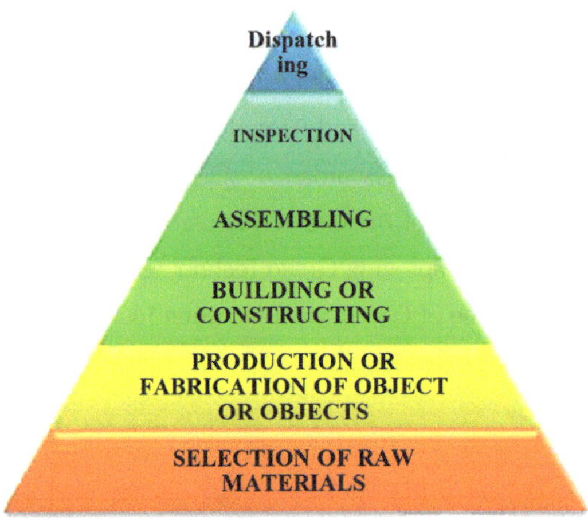

Figure 3.1: Sequence of manufacturing.

[Source: https://www.researchgate.net/publication/305731124_GREEN_MANUFACTURING_TECHNOLOGIES_-_A_REVIEW]

The word manufacturing is derived from Middle French in Medieval Latin *Manufactura* which means prepared by hand. From the Stone Age, several manufacturing methods have been developed (Connell, 2010). But the first symbols of modernized manufacturing technique adoption were comprehended between the 16th and 17th centuries. Manufacturing is the sequence of procedures comprising:

a) Selection of the raw materials.
b) Fabrication or production of objects or object.
c) Constructing or building.
d) Assembling the parts.
e) Examination or inspection and finally dispatching (Deif, 2011; Hwang & Huynh, 2015).

This is signified by the figure above. It is the obligation of industrialists to grow the economy of the country with modern manufacturing techniques and their products. Innovation is going more rapidly so as the necessities of humans. In order to fulfill their necessities, the Industrial sectors of several countries are adopting modern manufacturing techniques (Deliza et al., 1999; Watanabe, 1999; Xun, 2017). Along with the modern techniques, they should also see that existing or new method isn't damaging the natural resources and isn't troubling the chain of the ecosystem. Since the accessibility of natural resources isn't in surplus their use must be adjusted for future generations. The ingestion of energy and natural sources by humans is increasing exponentially. So frugal engineering is the need of an hour (Shrivastava, 1995; Dangelico & Pontrandolfo, 2010; Iravani et al., 2017).

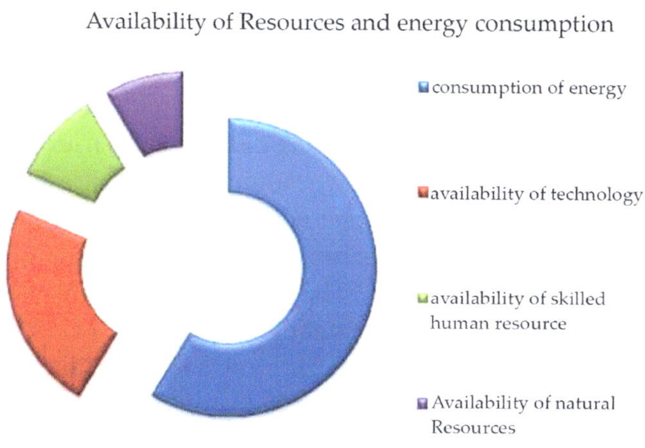

Figure 3.2: Comparison between energy consumption and availability of resources

[Source: https://www.researchgate.net/publication/305731124_GREEN_MANUFACTURING_TECHNOLOGIES_-_A_REVIEW]

3.2. CONVENTIONAL MANUFACTURING TECHNOLOGIES

There are several conventional manufacturing technologies like casting, molding, machining, joining, forming, rapid manufacturing, additive manufacturing, etc. In the majority of the circumstances, manufacturing is associated with the production industry (Tsai et al., 2013; Ahn, 2014).

Casting – A customary manufacturing procedure in which the metal in molten form is drizzled into the cavity of chosen shape and then left for cooling. After the metal gets hardened the mold is then taken out of the cavity. In this, there exists plaster casting, metal casting, plastic resin casting, and concrete casting. This practice is followed traditionally. Modern casting is normally divided into two categories non-expendable and expendable casting. Expendable casting is utilized for plastic, sand, wax and plaster moldings. In this technique, the molds are generally of temporary and non-reusable which upsurge the waste of manufacturing which is the non-biodegradable product (Zhou et al., 2015; 2018). In non-expendable casting, the mold needs to be transformed after every cycle. In this type of casting permanent mold casting, semi-solid metal casting, die casting, continuous casting and centrifugal casting are done. There are several other modern techniques of casting but the disposed waste disposed of the casting industry is the main criterion. This adds to contamination and has a huge harmful effect on the environment. The casting industry is contaminating the environment due to the nature of work (Dornfeld, 2014). Some of the environmental problems linked with the casting industries are the waste of foundry comprising of zinc, lead, cadmium which are harmful to the workers, discharge of dangerous air contaminants that comprise carbon materials, gaseous contaminant benzene which gets discharged from the cupola furnace is harmful to health. The environmental management system is developed for planning, implementation, inspection and acting model for persistent regulation of waste and inspection on the production. But the spread of this system is restricted to the large industries in which contemporary infrastructure is available. The levels of contaminant emissions are high in the small scale industries. A centralized system must be developed in order to control the level of air pollution (Azzone & Noci, 1998; Liu et al., 2017).

Machining – A procedure in which the metal block is cut into anticipated shape by precise metal elimination with the help of machines like lathe, driller, planner, slotter, etc. Machining comprises abrading, cutting, grinding, drilling, slotting and shaping. These procedures are done by different kinds of machining processes like ultrasonic machining, computer numerical machining, electrochemical machining, dry machining, photochemical machining, conventional machining, etc. Chip produced in machining procedure is usually wasted and it is a very time-consuming method. For the purpose of cutting, fluids like water-based – emulsions, chemical solutions, mineral oils, fatty oils, extreme pressure oils, composed oils, multiple oils are used. These metal cutting fluids cause contamination because they cannot be utilized for more cycles once the process is performed, these fluids are then thrown out into the soil and water bodies causing hazardous pollution. This must be diminished by a reduction in the volume of working fluids, utilizing substitute metalworking fluid application approaches, recycling technologies, etc. Additive manufacturing helps in less material loss and utilization of the cutting fluids is also reduced. But the discharge of toxins isn't decreased even in this manufacturing method (Florida, 1996). It does not offer any fair and quantitative inclusive life cycle assessment. Whether it is additive manufacturing or conventional machining the health issues and contamination caused by both of the techniques because of the working fluids are of the same extent. Hence there must be a continuous inspection of pollutant levels at every stage of manufacturing (Sangwan, 2006).

3.3. NEED FOR GREEN MANUFACTURING TECHNOLOGIES

The business would be quite fruitful if its emphasis on the developments in the global market and within the industry. Nowadays in any developing or developed country the term "Green Technology" is being reverberated in every industrial sector. It is mainly related to the production industry (Kreiger & Pearce, 2013; Klahn et al., 2014; Merklein et al., 2016). Because of the different kinds of production methods, there have been several greenhouse gases discharged into the atmosphere for the last hundred years (Emmelmann et al., 2011; Paris et al., 2016). This triggered extreme environmental phenomena of 'global warming.' Now it is the time to save the planet from the influence of these dangerous gases. Hence several eco-friendly operations and sustainable practices have been standardized in almost every industry. As there is a huge amount of energy ingestion in traditional manufacturing methods, to decrease that usage numerous industries adopt green technologies (Morrow et al., 2007; Salonitis & Al Zarban, 2015). Green technologies optimize reduce wastage, resource consumption, adopt recycling and recycling techniques so that the expense of production can be optimized (Nau et al., 2011; Mueller, 2012).

3.4. PREVENTION, CONTROL, AND REMEDIATION FOR CONTAMINATION

Conventionally, there are three methods of addressing pollution:

a) Prevention;
b) Control; and
c) Remediation.

These create the hierarchy, where the first option or priority is prevention, trailed by the control measures, with remediation as the third option. Pollution reduction can refer to the means that decreases pollution, or the mitigation of pollution; in practice, it generally means control (Sezen & Cankaya, 2013; Maruthi & Rashmi, 2015). Although the hierarchy of these three concepts is in terms of priority or preference, this isn't always so practically: there might be controlling pressures to select one path instead of another; one approach might be less costly than the another, or remediation might be the most crucial; for instance, in the case of the major spill or hazardous dissemination of contaminants from the contaminated site (Rusinko, 2007; Murugesan, 2008; Paul et al., 2014).

3.4.1. Pollution Prevention

Pollution prevention can normally be defined as the strategy which averts the creation of contaminants (Digalwar et al., 2013; Kong et al., 2016). In the phrase of Barry Commoner, "If it is not there, it cannot pollute." Therefore, if the chemical whose utilization outcomes in pollution are eradicated, there will be zero discharge of the pollutant. Zero discharge is more substantial if the chemical isn't substituted by another chemical – a substitute or alternative – which outcomes in the different pollutants (Mittal & Sangwan, 2014).

One central approach of pollution avoidance is banning, removal or the phasing out of particular chemicals or classes of the chemical. Such approaches are laid down in the shape of regulations or laws by the national governments, less frequent by international instruments or by the sub-national governments (Rew, 2007; Yongming, 2009).

The second approach is the reduction of pollution, in the context of avoidance rather than control. If the utilization of the chemical which results in contamination is decreased, then the outcome will always be less pollution. Pollution reduction approaches are exemplified in North America by TUR (toxics use reduction) programs and in Europe by clean technology programs.

Unlike phase-outs and bans, which normally apply to all workplaces within the political jurisdiction, pollution reduction programs apply to particular workplaces or classes of the workplace. These are generally industrial manufacturing workplaces over a certain size, in the first example, however, the principles of pollution lessening can be applied normally; for instance, to power plants, mines, offices, construction sites, agriculture, and municipalities. At least two United States states Michigan and Vermont have legislated toxic use reduction programs for the individual households which are workplaces.

Pollution reduction can outcome in the removal of particular chemicals, therefore accomplishing the same objectives as phase-outs and bans. Again, this would outcome in zero discharge of contaminant concerned, but necessities to remove specific chemicals aren't part of the pollution reduction programs; what is suggested is the general program with a supple range of stated methods. The necessity to remove the particular chemical is an instance of the "specification standard." The necessity to introduce the general program is a performance standard since it permits flexibility in the mode of execution, though a particular mandatory target for the general program would count as the specification standard. When the businessmen have to choose, they normally prefer performance to the specification standards (Commoner, 1994; Liu et al., 2006).

3.4.2. Pollution Control

Pollution control measures can't remove pollution; what these measures can do is to diminish the effects of pollution on the environment (Lober, 1998; Cabezas et al., 1999). Control measures are introduced at the end of the pipe." The helpfulness of control measures is dependent on the contaminant and the industrial condition (Shen, 1995; Bailey & Solomon, 2004).

The main approaches to pollution control are (Freeman et al., 1992; Sarkis & Cordeiro, 2001):

1. The capture and successive storage of pollutants.

2. Filtration, whereby waterborne or airborne pollutants are eliminated from the waste stream by physical approaches like filters, meshes, and the other permeable barriers.
3. Precipitation, whereby the contaminant is precipitated chemically and then seized in the transformed state or seized by physical methods like an electrostatic charge.
4. Destruction; for instance, neutralization or incineration, whereby contaminants are transformed biologically or chemically into the substances which are less dangerous.
5. Dilution, whereby the contaminant is flushed or diluted to reduce its effects on an ecosystem or on an organism; or concentration to decrease the effects of disposal.
6. Dissolution or evaporation; for instance, dissolving gas in the water.
7. Utilization; for instance, transforming a contaminant into a potentially useful product.
8. Out of the process of recycling.
9. Media-shift, whereby the waste-stream is distracted from one medium, like water, air, or soil, to another, on the basis that the medium-shift lessens the harmfulness of pollutants.
10. State-changes—a change of state to solid, liquid or gaseous on the basis that the new state will be less harmful.

3.4.3. Pollution Remediation

Remediation is required to the degree that pollution avoidance and control fail. It is also costly, with the expense not always growing to the polluter (Crawford, 1976; Downing & White, 1986). The modes of remediation are discussed below.

3.4.3.1. The Clean-Up of Polluted Sites

Clean-up usually has a simple meaning, as when the employer is needed to clean up his act, which can have numerous meanings. Within environmental protection, clean-up refers to technical terms meaning a mode or a branch of remediation. Even within this constrained utilization of the term, clean-up can generally mean

(1) Elimination of pollutants from the contaminated site.

(2) Rehabilitation of the site so that it is refurbished to its use-potential.

Again, clean-up occasionally refers to the containment of pollutants within the site, body or area of water—for instance, by sealing, capping, or the building of an impermeable floor (Montgomery, 1972; Segerson, 1988).

To be fruitful, clean-up must be 100 percent effective, with full safeguard for workers, the general public and the bystanders. A further concern is whether the clean-up methods, materials, and technology don't create further hazards. Though it is appropriate to utilize engineering controls to safeguard clean-up workers, there will always be a necessity for suitable personal protective equipment. Normally, workers involved in the remediation are categorized as hazardous-waste workers; however, features of such kind of work are usually undertaken by municipal workers and firefighters, among others.

A large number of chemical, physical, biotechnological and biological agents and approaches are utilized in clean-up of the contaminated sites.

3.4.3.2. Hazardous-Waste Treatment

Most treatment of dangerous waste now occurs in purpose-built facilities by dangerous-waste workers. From the viewpoint of environment, the test of efficiency of a harmful-waste facility is that it yields no outputs which aren't inert or virtually inert, like insoluble inorganic compounds, silica, insoluble and non-corrosive slags, gaseous carbon dioxide or nitrogen however carbon dioxide is the greenhouse gas which causes change in climate, and is, therefore, an environmental detriment (Banat, 1995; Esterkin et al., 2005).

An additional test is that the workplace is energy efficient – that is, energy isn't wasted and as energy non-intensive as feasible. The general rule of thumb is that the more efficient the pollution abatement approach, the more energy is ingested, which by the criteria of sustainable development is another detriment (Sirés & Brillas, 2012; Wu et al., 2017).

Even when the employers are appropriately protected, it is quite easy to observe the shortcomings of dangerous-waste treatment as the mode of addressing contamination. Pollution prevention approaches can be applied to the operation of the treatment procedures but they can't be made practical to the principal input – the waste to be treated. The facilities of Hazardous-waste treatment will usually need as a minimum as much energy in order to treat the wastes as was used in its formation and there will be an extra waste as an output, though non-toxic or inert (Vaseashta et al., 2007; Mamba & Mishra, 2016).

3.4.3.3. Spills and Leaks

The same concerns will apply to the chemical leaks and spills as to the clean-up of the contaminated sites, with the surplus hazards caused by the urgency of the clean-up. Workers cleaning up leaks and spills are almost emergency workers. Depending on the nature and the scale of the pollutant, spills and leaks can become chief industrial accidents (Von Wandruszka, 2000; Valls & De Lorenzo, 2002).

3.5. MODES OF POLLUTION AVOIDANCE

3.5.1. Definition and Philosophy

The definition of pollution avoidance or prevention might seem to be an unimportant matter, but it is vital as advocates of the pollution prevention desire, as the principle of policy, to witness a single-minded and hostile prevention approach at the cost of control techniques, and to avert remediation. The more strictly prevention of pollution is defined; the more probable it is to prosper as a practical approach. Conversely, the more extensively employers are permitted to define this term, the more probable their activities are to outcome in a mixture of the same failed strategies. Employers occasionally reply that even the toxic wastes can possess a market value and control techniques to have their place, thus pollution is only potential pollution. Moreover, zero discharge is not possible and triggers misguided strategies and false expectations.

Most of the stringent definitions of the pollution prevention have, as a central element or sole, the prevention of the utilization of chemicals which outcome in pollutants so that the pollution isn't created.

3.5.2. Objectives

One probable aim of pollution avoidance is zero discharge of the pollutants. This is occasionally referred to as virtual elimination, as even zero discharge can't solve the problem of pollutants already existent in the environment. Zero discharge of the pollutants is possible utilizing pollution prevention methods. For example, one can imagine automobile manufacturing in which there exists zero discharge of contaminants from the plant; the other waste is reprocessed and the product comprises parts that are recyclable or reusable. Certainly, zero discharge of particular pollutants has been accomplished; for instance, by adjusting the production procedure in the wood pulp mills so that no furans or dioxins are released in the effluent. The objective of zero discharge has been written into the environmental laws and into policies of bodies appointed to abate pollution (Dvorak et al., 2008; Youngblood et al., 2008).

Practically, zero discharge frequently gives way to aim reductions; for instance, a 50 percent reduction in the pollution emissions by that year. These interim targets or targets are generally in the form of aims or challenges by which to calculate the accomplishment of to pollution prevention program. They are hardly the product of a possibility calculation or analysis and there are consistently no drawbacks attached to failure in order to attain the target. Nor are they calculated with any precision (Koutalakis et al., 2010; Galeazzo et al., 2014).

Reductions must be measured by variations in the formula:

Pollution = Toxicity of the pollutant × Volume of the discharges

or:

$P = T \times V \times E$ (exposure potential).

This is quite problematic in theory and costly in practice, however, it could be completed in principle by using hazard assessment techniques. The whole issue recommends that the resources would be better assigned elsewhere; for instance, in confirming that appropriate pollution prevention plans are made.

3.5.3. Methods

The key methods of pollution avoidance are:

i The phasing out or elimination of particularly hazardous chemicals.

ii Input substitution – substituting a hazardous or toxic substance with the less hazardous or non-toxic substance or with the non-toxic process. Examples are the replacement of the water-based synthetic organic dyes in the printing industry; citrus or water-based solvents for the organic solvents and in some applications, the replacement of vegetables for the mineral oils. Examples of the non-chemical substitution comprise the replacement of pellet blasting expertise for utilization of the fluid chemical paint strippers; utilization of the high-pressure hot water systems as a replacement for caustic cleaning and substitution of the kiln-drying for utilization of PCPs (pentachlorophenol) in the lumber industry. In all circumstances, it is essential to perform a replacement analysis to confirm that alternatives are genuinely less dangerous. This is a matter of structured common sense and at best the use of hazard assessment techniques to the chemical and proposed substitute.

iii Product reformulation – replacing for a prevailing end-product which is less hazardous or non-toxic upon utilization, disposal or release. While input replacement refers to the adjuncts and raw materials at the front end of the production procedure, product reformulation methods the problem from the final product of the production cycle.

General programs to yield products that are environmentally benign are instances of economic conversion. Examples of specific measures in the area of the product reformulation comprise the making of rechargeable batteries as an alternative of the throw-away kinds and the utilization of water-based coatings instead of the ones based on the organic solvents. Again, substitution analysis is essential to confirm that the net environmental advantage is greater for reformulated products than for the originals (De Bruijn & Hofman, 2002; Liu et al., 2016).

i. Production unit redesign modification or modernization, which outcomes in less chemical utilization or in the utilization of less toxic substances.
ii. Enhanced operation and maintenance of production methods and production, comprising better housekeeping, more effective production quality control and procedure inspections. Examples are the measures of spill prevention, the utilization of spill-proof containers, leak prevention and floating lids for the solvent tanks.
iii. Using less and reprocessing more. For example, some degreasing operations occur too often on a single item. In other circumstances, chemicals can be utilized more sparingly in every operation. De-icing fluids can occasionally be reused, a circumstance of extended use.
iv. Closed-loop approaches and in-process recycling. Strictly speaking, the closed-loop procedure is one in which there aren't any emissions into the environment or workplace, not even

wastewater into the surface water or CO_2 into the atmosphere. There are only finished products, inputs, and non-toxic or inert wastes. Practically, closed-loop methods remove some but not all of the hazardous releases.

3.5.4. Recycling

Any definition of pollution prevention is possible to outcome in the number of grey areas in which it isn't easy to differentiate prevention measures from the emission controls. For example, to qualify as the prevention method, a phase of the production process might have to be an integral part of production unit, but how far the phase must be from the border of production process to qualify as the prevention measure isn't always clear (Hilaly & Sikdar, 1996; Petraru & Gavrilescu, 2010). Some processes might be so isolated from the heart of operation that these procedures look more like an add on the procedure and, therefore, more like an end of pipe control measure than the prevention method. There are unclear circumstances like the waste pipe that offers the feedstock for the neighboring plant: taken together, the plants deliver a kind of the closed-loop but the upstream plant still yields effluent and, therefore, fails the prevention test (Green et al., 1983; Abraham, 2000).

Conventionally, there are three types of recycling:

1. In-process recycling; for instance, when the solvent of dry-cleaning is filtered, washed and dried, then reutilized within the single process.
2. Out-of-process but on the site, as when the production waste of pesticide is washed and then reutilized as the inert base in the new production run.
3. Out-of-process and off the site.

Of all the above, the third is generally ruled out as not succeeding for pollution prevention: the more faraway the site of recycling, the less of the assurance that the recycled product is reused. There are also dangers in transporting the wastes to be recycled and the fiscal doubt that the waste will normally have a consistent market value. Similar, although less acute, concerns apply to the out-of-process but on the site of recycling: there is the likelihood that the waste won't be recycled and if recycled, not essentially reused (Freeman et al., 1992; Cagno et al., 2005).

In the primary pollution prevention approaches of the 1980s, out-of-process but on the site, recycling was not included as not being the genuine pollution prevention measure. There was anxiety that an efficient pollution avoidance program would be diluted or compromised by too great a focus on recycling. In the mid-1990s, some of the policy-makers were ready to entertain out-of-process but on-site recycling as the genuine pollution prevention method (Knight, 1995). One aim was that there are genuine grey areas between control and prevention. Another reason was that some on the site recycling actually does what it is meant to do, although it might not qualify technically as pollution prevention. The third aim is business pressure: employers witness no reason why approaches should be ruled out if they serve the objectives of the pollution prevention program (Sullivan & Floyd, 1991; El-Halwagi, 1998).

3.5.5. Pollution Prevention Planning

Planning is a necessary part of the pollution prevention methodology because the advantages in both environmental protection and industrial efficiency are possibly to be in the longer-term, outlining the sort of planning that usually goes into product design and marketing. The production of interrupted pollution avoidance plans is the usual way of comprehending pollution prevention planning. There isn't any model for such kinds of plans (Campbell et al., 1982; Sturgill et al., 2000). One proposal predicts:

1. Objectives and aims.
2. Chemical inventories and approximations of releases into the environment.
3. Pollution prevention methods utilized and methods proposed.
4. Action and duties in the event of a plan not being realized or fulfilled.

Another proposal predicts:

1. A review of the production processes.
2. Identification of pollution prevention chances.
3. A ranking of chances and a plan for the execution of the designated options.
4. Measures of the achievement of the plan after the execution period.

The prominence of such plans changes widely. Some are intentional, though they can usually be spelled out in the law as a code of practice. Others are essential in that they are needed

1. To be kept on the site for examination.
2. Submitted to the controlling authority on completion.
3. Submitted to the controlling authority for some kind of approval or scrutiny.

There are also changes, such as needing a plan in the event that the voluntary plan is in some way inadequate or ineffective (Şchiopu et al., 2007).

The extent to which essential plans are rigid also changes; for instance, in regard to sanctions and penalties. Only some of the authorities have the authority to require particular changes in the content of the pollution prevention plans; nearly all have the authority to require variations in a plan in the event that the official requirements haven't been met; for instance, if some of the plan headings haven't been addressed. There are almost no examples of sanctions or penalties in the event that the fundamental requirements of the plan haven't been met. In other words, legitimate requirements for pollution prevention planning are far away from traditional (Chang & Chung, 2000).

3.5.6. Legislation

In the Canadian provinces of Ontario and British Columbia, the measures of pollution prevention are voluntary; their effectiveness is dependent on moral suasion on the part of environmentalists and governments. In the U.S, about half of the states do have some kind of legislation, whereas in Europe, many northern countries have constituted clean technology programs. There is an extensive variety of both the effectiveness and content of such legislation. Some of the laws strictly define pollution prevention; others define it loosely or widely and cover a broad variety of environmental protection activities regarding waste and pollution, not just pollution prevention. The law of New Jersey is highly rigid; those of Commonwealth of the Massachusetts and States of Oregon and Minnesota comprise a high extent of government assistance and scrutiny and assistance.

3.5.7. Safety, Health, and Employment

Pollution avoidance is of central anxiety to the occupational health: if the utilization of toxic substances reduces, there will always be a conforming reduction in worker exposure to the toxic substances and, therefore, in industrial diseases. This is a major cause of prevention at the source of the dangers and, in many situations, the removal of dangers by engineering controls, the first and the best line of defense against chemical hazards. However, such measures of prevention are quite different from one customary approach, which is the total isolation or total enclosure of the chemical process. Whereas total enclosure is highly desirable and highly useable, it doesn't count as the pollution prevention approach since it controls, instead of that decreases intrinsically, a prevailing hazard (Quinlan et al., 2001; Underhill & Quinlan, 2011).

The contaminants which pose dangers to communities, workers, and the environment alike, have normally been addressed mainly because of their influence on human communities. Though the utmost exposures are frequently received by the workers within a workplace, this hasn't, so far, been the main focus of pollution prevention measures. The Massachusetts legislation, for example, targets to decrease the hazards to the health of consumers, workers, and the environment without comprising the risks between consumers, workers, and parts of the environment. But there wasn't any effort to emphasize workplace pollution as the major harm, nor was there any necessity to accord importance to the main human exposures to dangers – frequently the workers. Nor is there any obligation to train the workers in the discipline of pollution prevention (Harris, 2004; Lamm, 2010).

There are many reasons for this. The first reason is that pollution avoidance is comparatively a new discipline in the context of the general, customary failure to witness environmental protection as the function of procedures adopted and used within workplaces. The second reason is that the worker-management co-determination in the area of environmental protection isn't well advanced. Workers in several countries possess legal rights, for example, to join workplace safety and health committees; to refuse unhealthy or unsafe work; to safety and health information; and to get training in safety and health issues and procedures. But there are some legal rights in parallel and frequently overlapping area of the environmental protection, like the right to join union-management environment committees; right of the employees to blow the whistle on the anti-environmental practices of employer; the right to reject to degrade or to pollute the outside environment; right to the environmental information and the right to take part in the workplace environmental audits.

The influences of pollution avoidance planning on employment are quite hard to gauge. The explicit objective of pollution avoidance initiatives is frequently to increase environmental protection and industrial efficiency simultaneously and by a similar set of measures. When this occurs, the normal effect is to reduce overall employment within the given workplace but to upsurge the skills needed and then to upsurge job security. To the degree that the utilization of adjuncts and raw materials is decreased, there will be reduced chemical manufacturing employment, although this is probable to be offset by an implied transition of the feedstock to special chemicals and by the development of substitutes and alternatives (Evans & Repper, 2000; Bohle et al., 2004).

There is one feature of employment that pollution prevention planning can't address. Pollution emissions from the single facility might decrease but to the degree that there exists an industrial approach to create value-added employment and wealth, an upsurge in a number of the production facilities will incline to abolish the environmental protection advantages already achieved. The most tarnished failing in the environmental protection measures – that the pollution emission decreases and controls are abolished by an upsurge in a number of the sources applied, unluckily, to pollution prevention along with any other form of interference.

3.5.8. Workplace Environmental Audits

Pollution avoidance planning can be part of or be lodged in the workplace environmental audit. Though there are several versions of such kind of audits, they are possible to be in the shape of a site audit or production audit, where the entire production cycle is exposed to both a financial and environmental analysis.

There are roughly three areas of environmental protection and sustainable development which can be covered in the workplace audit:

1. Conservation of the natural resource inputs; for instance, water, minerals, and wood products.
2. Energy utilization, which might also comprise consideration of the energy sources, energy intensiveness, energy efficiency, and energy conservation.
3. Pollution prevention, control, and remediation.

To the degree that the pollution prevention is effective, there will be a conforming decrease in the significance of control and remediation measures; the pollution avoidance measures can form the main part of the workplace environmental audit.

Conventionally, businesses were able to externalize environmental disadvantages through means as the wasteful utilization of water or unloading the wastes onto the environment and outside the community. This has triggered demands for the taxes on the front end like water utilization or on outputs like environmentally unfavorable products or on the wastes.

In this way, expenses for the business are internalized. However, it has demonstrated hard to put the accurate price on inputs and on the disadvantages; for instance, the expense to an environment of wastes and communities. Nor is it quite clear that pollution taxes decrease pollution in ratio to the amounts imposed; taxes might well internalize expenses, but they otherwise add to the price of doing business.

The benefit of the environmental auditing is that the audits can normally make some economic sense without having to cost externalities. For example, the value of wastes can be estimated in terms of energy non-utilization and resource input loss. Unfortunately, the financial aspect of pollution avoidance planning and its role in workplace environmental audits isn't well advanced.

3.5.9. Hazard assessment

Some schemes of pollution prevention work without any danger assessment – that is, without conditions to decide whether the facility or plant is less or more environmentally gentle as an outcome of the pollution prevention measures. Such schemes might rely on the chemicals which are objects of consideration or which describe the opportunity of the pollution prevention program. But the list of chemicals doesn't score chemicals as to their comparative dangerousness, nor is there an assurance that the chemical replace not on the list is less hazardous than the listed chemical (Aleotti & Chowdhury, 1999).

Other schemes rely on the criteria for evaluating dangerousness, that is, on the hazard assessment systems. They work, fundamentally, by laying down the number of environmental factors, like bioaccumulation and persistence in the environment and the number of human health factors which assist as measures of poisonousness; for instance, carcinogenicity, acute toxicity, reproductive toxicity, mutagenicity and so on (Guzzetti et al., 2005).

There is then the weighted scoring system and decision process for scoring those factors on which there is insufficient information on the list of chemicals to be scored. Appropriate chemicals are then marked and ranked, then accumulated in groups in the descending order of dangerousness (Henschel et al., 1997; Huabin et al., 2005).

However such schemes are occasionally devised with a particular aim in mind; for instance, for evaluating priorities for elimination or for control measures – their essential utilization is as a theoretical scheme which can be utilized for a large range of the environmental protection measures, comprising pollution prevention. For example, the top group of marked chemicals could be the main candidates for the obligatory pollution prevention program or they might be candidates for substitution or phasing-out. Such schemes don't tell how much one should decrease environmental health dangers; they only tell that any measures that are taken must be informed by the risk assessment scheme (McGrath et al., 2004).

For example, if decisions are too made about replacing a less dangerous chemical for the more hazardous one, a scheme can be used to tell that whether, prima facie, the replacement decision is good or not: both of the chemicals can be run through the scheme in order to conclude whether there is a narrow or wide gap amongst the chemicals regarding their hazardousness. There are two kinds of considerations which hardly fall within the range of danger assessment schemes. The first one is exposure data or the possibility of human exposure to the chemical. The second is hard to calculate and arguably it changes the intrinsic hazard of the concerned chemicals. For example, a chemical could be given a low urgency on the ground that the potential of its exposure is low; though it might, in fact, be comparatively easy to deal and highly toxic (Andersen, 1966; Ashby et al., 1994).

The second kind of consideration is the socioeconomic influence of removing or reducing the utilization of the concerned chemical. While the substitution decisions can be made on the basis of hazard analysis, one would have to make distinct socioeconomic analysis and consider, for instance, the social helpfulness of the product linked with the chemical utilization and one would also have to keep in mind the influence on workers and the communities. The objective for keeping such kind of analysis distinct is that it is unbearable to score the outcomes of the socio-economic analysis in a similar manner that the intrinsic dangers of the chemicals are scored. There are two totally separate sets of values with different rationales (Landrum et al., 1992).

However, danger assessment schemes are critical in assessing the accomplishment of pollution prevention programs. For example, it is likely to apply the schemes without any reference to risk analysis, risk assessments, and without reference to the cost-benefit analysis. An earlier strategy to pollution was to perform risk assessment first and only then determine what kind of action and how much was essential to lessen the risk to an acceptable level. The outcomes were rarely dramatic. Hazard assessment, alternatively, can be used very rapidly and in a way that it doesn't compromise or delay the effectiveness of pollution prevention programs. Pollution prevention is a realistic program capable of speedily and constantly addressing pollution problems before they arise and as they arise.

3.6. DIFFERENT TYPES OF THE GREEN MANUFACTURING TECHNOLOGIES

1. *Operation of the energy audits* – The main step for the industry in order to meet the ecological status is to perform an energy audit on an everyday basis and to witness where the energy is being used in large amounts and substitute it with the green-technical equipment. For instance, substituting traditional light bulbs with CFLs (Compact Fluorescent Light Bulbs). The variation in consumption does not get reported at once but in the long run, there will be variation in energy consumption. As there is a decrease in the availability of fossil fuels and global ecological balance is being disturbed, henceforth the only survival approach is to reduce energy consumption. This holds for the individual as well as for the industry. Green products, which are exemplified with less consumption of energy, when engaged in production and manufacturing would improve the decrease in carbon emissions (Azzone & Noci, 1998; Bartlett & Trifilova, 2010).
2. *Use of renewable and substitute energy* – Shifting to the usage of the conventional power which is made by fossil fuels to the power produced by wind, tides, solar, biomass, etc. is the necessity of the hour. It is a better option of building their own eco-power source on the site so that necessities are met and the emissions are reduced (Chuang & Yang, 2014). Several governments offer tax-free production and incentives for the production of the power from green technologies. If it isn't possible for the building such kind of power sources, these can be bought from the external source. The costs experienced for the building of green power sources might be high but when compared to the ecological advantages and optimization of energy this will be a huge success

in the long term (Sezen & Cankaya, 2013; Liu et al., 2017).

3. *Application of three R's (Reduce, Reuse and Recycle)* – Another method to decrease the cost of production is to recycle the production waste and building of the recycling plants so that waste removal expenses would be reduced.
4. *Minimize the consumption of water* – Numerous industries add to water pollution – chiefly production, leather, textile, and construction. Its influence is felt directly on human beings and the other creatures.
5. Choosing substitute manufacturing techniques.
6. Consumption of telecommuting.
7. Consumption electronic mails instead of paper mails.
8. Shifting to hybrid vehicles.

3.7. COMPANIES EMPLOYING GREEN TECHNOLOGIES

1. The leading industrial product of the world is steel. Besides its production, the greenhouse gases discharged from conventional manufacturing methods are also high. Around 1.5 billion tons of steel is being produced annually throughout the world. Each ton of steel production discharges two tons of CO_2 into the atmosphere. This constitutes nearly 5 percent of global warming and pollution. The researchers of MIT (Massachusetts Institute of Technology) have established a technology that decreases carbon emissions. Material Chemistry professor Donald Sadoway discovered a new procedure of yielding highly pure steel with almost 0 percent carbon emission – in its place the oxygen is produced. The process consists of utilizing Iridium anode that utilizes Iron oxide in order to yield oxygen in abundance. This process is recommended by Donald Sadoway during his research work of yielding oxygen on the lunar soil with the aid of NASA's grant. As the anode utilized is Iridium which is very costly makes the production of the steel costly in small industries. It would be feasible and reliable for production on large scales. But finding the substitute of such metal having a high melting point around 1723^0C is difficult.

2. The main objective of any production procedure is to finish manufacturing with less influence on the environment, uses optimal energy for its production and yield minimal waste. Adopting the green manufacturing method has a direct influence on the environment – by yielding secondary products from the main source. For instance in Japan Kao's company – by utilizing energy-saving methods which directly decreases the emission of greenhouse gases. One of the methods adopted by them is the cogeneration – combined power and heat which is engaged in plants that yields electricity from main sources at a temperature of around 500°C and is reused as the input into boilers.
3. Around 80 percent of the main source is transformed into the service or product. The prime objective of Koa's is to reduce CO_2 emissions. Kao's started utilizing heat pumps which are utilized for air conditioning and refrigerating. Heat pumps have the capacity to transfer heat from hot to cold bodies and vice versa. From the 1970's they began shifting to renewable energy sources from fossil fuels.
4. Production of the solar photovoltaic cell, which is chiefly made up of silicon, by the polysilicon manufacturing technique relates to the handling of the hazardous gases and consumption of high power of around 120 KW for producing 1 kg of silicon. This process chiefly handles trichlorosilane gas which is dangerous to the surrounding atmosphere and also for workers. The silicon process does not require handling such kind of gases and yields the same amount of the silicon as that mentioned above with around 20–30 KW of power. The process consists of melting the metallurgical grade silicon which has contaminations of phosphorus and boron in the aluminum smelt of almost 1023°K temperature. The silicon achieves almost 97 percent of purity during this step. After this step, the silicon forms into the flakes and the contaminations stay in aluminum, on cooling the flakes of silicon get solidified and the impurities of aluminum are drained out, which can be utilized in purification of the water. Acid is then poured onto the flakes of silicon so that the rest of the aluminum layer can be eliminated by just draining. There will be a thin layer of aluminum-coated over silicon finally which can be eliminated by simple methods.

5. The construction industry is one of the main sources of pollution and discharges 4 percent of particulate contaminants into the atmosphere. It contributes to water pollution, noise pollution, and air pollution. Construction of ecological structure won't only help in providing an improved outdoor atmosphere but also a better indoor atmosphere. Conventional construction structures and materials have triggered numerous health problems. Along with biological contaminants like dust mites, the chemical contaminants from paints, solvents, plastics, varnishes, and composite timber are the major reasons for health problems like headaches, depression, asthma, eczema, chronic fatigue syndrome, and palpitations. They have even tertiary and secondary effects. Green buildings made of non-toxic and natural products remove several of these problems. For over 5,000 years lime was utilized as the building material. The modern building materials comprise cement. In the UK several ancient buildings are made of lime. Lime can be utilized in several forms like lime putty; limewash and lime mortar for construction purposes. When lime is used in the plaster form it absorbs CO_2 from the surroundings and forms the thick calcium carbonate layer. It permits the moisture from the inside surface to the outside, which is done in the circumstance of contemporary paintings on the buildings with toxic gas liberation. It also traps the required moisture.
6. Some multi-national companies are asking for to go green. IKEA (Ingvar Kamprad Elmtaryd Agunnaryd) is utilizing timber for producing affordably priced furniture. It also amplified its investments in renewable energy. Nike, Philips Electronics, Johnson & Johnson, Earth Tec, Tupperware, and Dell are some of the companies which are following the going green technologies.

3.8. THE OUTCOME OF GREEN MANUFACTURING TECHNOLOGIES

The diagram below signifies the development of industry and therefore the country when it accepts green manufacturing technologies. An industry or country is said to be advanced not with the highest modern manufacturing technologies or production rates, but when it is fruitful at the production rates with the implementation of eco-friendly manufacturing technologies (Xiaohui, 1999; Fei, 2000). The development is accomplished stepwise.

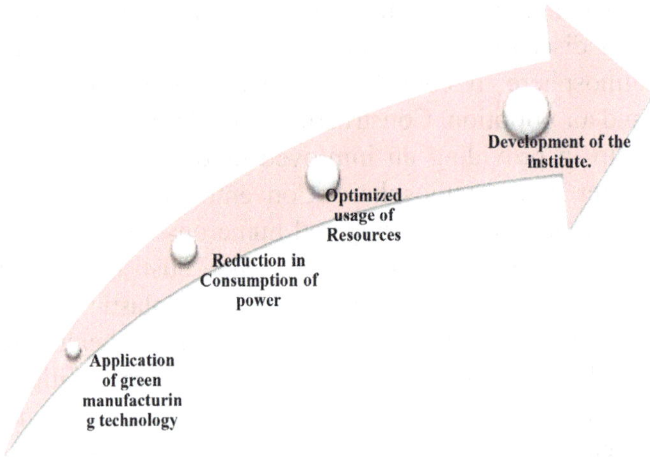

Figure 3.3: Result of executing Green Technologies.

[Source: https://www.researchgate.net/publication/305731124_GREEN_MANUFACTURING_TECHNOLOGIES_-_A_REVIEW]

REFERENCES

1. Abraham, M. A. (2000). Pollution prevention course that helps meet EC 2000 objectives. *CHEM ENG EDUC*, *34*(3), 272–277.
2. Ahn, S. H. (2014). An evaluation of green manufacturing technologies based on research databases. *International Journal of Precision Engineering and Manufacturing-Green Technology*, *1*(1), 5–9.
3. Aleotti, P., & Chowdhury, R. (1999). Landslide hazard assessment: summary review and new perspectives. *Bulletin of Engineering Geology and the Environment*, *58*(1), 21–44.
4. Andersen, A. A. (1966). A sampler for respiratory health hazard assessment. *American Industrial Hygiene Association Journal*, *27*(2), 160–165.
5. Ashby, J., Brady, A., Elcombe, C. R., Elliott, B. M., Ishmael, J., Odum, J., ... & Purchase, I. F. H. (1994). Mechanistically-based human hazard assessment of peroxisome proliferator-induced hepatocarcinogenesis. *Human & Experimental Toxicology*, *13*(2_suppl), S1–S117.
6. Azzone, G., & Noci, G. (1998). Identifying effective PMSs for the deployment of "green" manufacturing strategies. *International Journal of Operations & Production Management*, *18*(4), 308–335.
7. Bailey, D., & Solomon, G. (2004). Pollution prevention at ports: clearing the air. *Environmental Impact Assessment Review*, *24*(7–8), 749–774.
8. Banat, I. M. (1995). Biosurfactants production and possible uses in microbial enhanced oil recovery and oil pollution remediation: a review. *Bioresource Technology*, *51*(1), 1–12.
9. Bartlett, D., & Trifilova, A. (2010). Green technology and eco-innovation: Seven case-studies from a Russian manufacturing context. *Journal of Manufacturing Technology Management*, *21*(8), 910–929.
10. Benach, J., & Muntaner, C. (2007). Precarious employment and health: developing a research agenda. *Journal of Epidemiology & Community Health*, *61*(4), 276–277.
11. Bohle, P., Quinlan, M., Kennedy, D., & Williamson, A. (2004). Working hours, work-life conflict and health in precarious and" permanent" employment. *Revista de Saúde Pública*, *38*, 19–25.
12. Cabezas, H., Bare, J. C., & Mallick, S. K. (1999). Pollution prevention with chemical process simulators: the generalized waste reduction (WAR) algorithm—full version. *Computers & Chemical Engineering*,

23(4–5), 623–634.

13. Cagno, E., Trucco, P., & Tardini, L. (2005). Cleaner production and profitability: analysis of 134 industrial pollution prevention (P2) project reports. *Journal of Cleaner Production, 13*(6), 593–605.
14. Campbell, M. E., Glenn, W. M., & Pimm, L. R. (1982). Profit from pollution prevention: A guide to industrial waste reduction and recycling. *Pollution Probe,* 12 Madison Ave., Toronto, ONT., Canada M 5 R 2 S 1. 1982.
15. Chang, I. S., & Chung, C. M. (2000). Pollution prevention for manufacturing of ammonium chloride-an experimental study of wastewater recycling. *Desalination, 127*(2), 145–153.
16. Chuang, S. P., & Yang, C. L. (2014). Key success factors when implementing a green-manufacturing system. *Production Planning & Control, 25*(11), 923–937.
17. Commoner, B. (1994). Pollution prevention: putting comparative risk assessment in its place. *Worst Things First? The Debate Over Risk-Based National Environmental Priorities*, 1, 203–208.
18. Connell, K. Y. H. (2010). Internal and external barriers to eco-conscious apparel acquisition. *International Journal of Consumer Studies, 34*(3), 279–286.
19. Crawford, M. (1976). *Air pollution control theory* (Vol. 1, p. 259). New York: McGraw-Hill.
20. Dangelico, R. M., & Pontrandolfo, P. (2010). From green product definitions and classifications to the Green Option Matrix. *Journal of Cleaner Production, 18*(16–17), 1608–1628.
21. De Bruijn, T., & Hofman, P. S. (2002). Partners in Pollution Prevention. In *Partnership and Leadership* (Vol. 1, pp. 195–215). Springer, Dordrecht.
22. Deif, A. M. (2011). A system model for green manufacturing. *Journal of Cleaner Production, 19*(14), 1553–1559.
23. Deliza, R., Rosenthal, A., Hedderley, D., MacFie, H. J. H., & Frewer, L. J. (1999). The importance of brand, product information and manufacturing process in the development of novel environmentally friendly vegetable oils. *Journal of International Food & Agribusiness Marketing, 10*(3), 67–77.

24. Dornfeld, D. A. (2014). Moving towards green and sustainable manufacturing. *International Journal of Precision Engineering and Manufacturing-Green Technology, 1*(1), 63–66.
25. Downing, P. B., & White, L. J. (1986). Innovation in pollution control. *Journal of Environmental Economics and Management, 13*(1), 18–29.
26. Dvorak, B. I., Hygnstrom, J. R., Youngblood, D. J., Woldt, W. E., & Hawkey, S. A. (2008). Lessons learned concerning impact assessment: pollution prevention technical assistance in Nebraska. *Journal of Cleaner Production, 16*(6), 751–760.
27. El-Halwagi, M. M. (1998). Pollution prevention through process integration. *Clean Products and Processes, 1*(1), 5–19.
28. Emmelmann, C., Sander, P., Kranz, J., & Wycisk, E. (2011). Laser additive manufacturing and bionics: redefining lightweight design. *Physics Procedia, 12*, 364–368.
29. Esterkin, C. R., Negro, A. C., Alfano, O. M., & Cassano, A. E. (2005). Air pollution remediation in a fixed bed photocatalytic reactor coated with TiO2. *AIChE Journal, 51*(8), 2298–2310.
30. Evans, J., & Repper, J. (2000). Employment, social inclusion and mental health. *Journal of psychiatric and mental health nursing, 7*(1), 15–24.
31. Fei, L. (2000). Green changes and innovations of the manufacturing industries in the twenty-first century. *Chinese Journal of Mechanical Engineering, 36*(1), 7–10.
32. Florida, R. (1996). Lean and green: the move to environmentally conscious manufacturing. *California Management Review, 39*(1), 80–105.
33. Freeman, H., Harten, T., Springer, J., Randall, P., Curran, M. A., & Stone, K. (1992). Industrial pollution prevention! A critical review. *Journal of the Air & Waste Management Association, 42*(5), 618–656.
34. Galeazzo, A., Furlan, A., & Vinelli, A. (2014). Lean and green in action: interdependencies and performance of pollution prevention projects. *Journal of Cleaner Production, 85*, 191–200.
35. Green, L. W., Wilson, R. W., & Bauer, K. G. (1983). Data requirements to measure progress on the objectives for the nation in health promotion and disease prevention. *American Journal of Public Health, 73*(1), 18–24.

36. Guzzetti, F., Reichenbach, P., Cardinali, M., Galli, M., & Ardizzone, F. (2005). Probabilistic landslide hazard assessment at the basin scale. *Geomorphology*, *72*(1–4), 272–299.
37. Harris, P. (2004). From health and safety to employee participation? The impact of the New Zealand Health and Safety in Employment Amendment Act (2002). *International Employment Relations Review*, *10*(1), 1.
38. Henschel, K. P., Wenzel, A., Diedrich, M., & Fliedner, A. (1997). Environmental hazard assessment of pharmaceuticals. *Regulatory toxicology and Pharmacology*, *25*(3), 220–225.
39. Hilaly, A. K., & Sikdar, S. K. (1996). Process simulation tools for pollution prevention. *Chemical Engineering*, *103*(2), 2–30.
40. Huabin, W., Gangjun, L., Weiya, X., & Gonghui, W. (2005). GIS-based landslide hazard assessment: an overview. *Progress in Physical Geography*, *29*(4), 548–567.
41. Hwang, C. L., & Huynh, T. P. (2015). Investigation into the use of unground rice husk ash to produce eco-friendly construction bricks. *Construction and Building Materials*, *93*, 335–341.
42. Iravani, A., Akbari, M. H., & Zohoori, M. (2017). Advantages and disadvantages of green technology; goals, challenges and strengths. *Int J Sci Eng Appl*, *6*(9), 272–284.
43. K. Digalwar, A., R. Tagalpallewar, A., & K. Sunnapwar, V. (2013). Green manufacturing performance measures: an empirical investigation from Indian manufacturing industries. *Measuring Business Excellence*, *17*(4), 59–75.
44. Kakuta, M., Tanaka, H., Sato, J., & Noguchi, K. (1981). A new calcining technology for manufacturing of coke with lower thermal expansion coefficient. *Carbon*, *19*(5), 347–352.
45. Karana, E., Hekkert, P., & Kandachar, P. (2009). Meanings of materials through sensorial properties and manufacturing processes. *Materials & Design*, *30*(7), 2778–2784.
46. Klahn, C., Leutenecker, B., & Meboldt, M. (2014). Design for additive manufacturing–supporting the substitution of components in series products. *Procedia Cirp*, *21*, 138–143.
47. Knight, C. F. (1995). Pollution prevention, technology challenges, and competitive advantage in the process industries. *Environmental Quality Management*, *5*(1), 87–92.

48. Kong, T., Feng, T., & Ye, C. (2016). Advanced manufacturing technologies and green innovation: The role of internal environmental collaboration. *Sustainability*, *8*(10), 1056.
49. Koutalakis, C., Buzogany, A., & Börzel, T. A. (2010). When soft regulation is not enough: The integrated pollution prevention and control directive of the European Union. *Regulation & Governance*, *4*(3), 329–344.
50. Kreiger, M., & Pearce, J. M. (2013). Environmental life cycle analysis of distributed three-dimensional printing and conventional manufacturing of polymer products. *ACS Sustainable Chemistry & Engineering*, *1*(12), 1511–1519.
51. Kumar, R., & Rao, P. S. (2018). Green Manufacturing Technology-Solution for Environmental Impact and Waste. *Sustainable development*, *4*(12), 1–21.
52. Lamm, Felicity (2010). Participative and productive employment relations: the role of health and safety committees and worker representation. *Employment Relationships. Workers, Unions and Employers in New Zealand*, 1, 168–184.
53. Landrum, P. F., Lydy, M. J., & Lee, H. (1992). Toxicokinetics in aquatic systems: model comparisons and use in hazard assessment. *Environmental Toxicology and Chemistry*, *11*(12), 1709–1725.
54. Liu, F. J., Gao, H. Y., & Wu, J. (2006). Transfer of cadmium in food chain and its prevention and control from pollution. *J. Agro-Environ. Sci*, *25*, 805–809.
55. Liu, G., Yang, Z., Chen, B., Zhang, Y., Su, M., & Ulgiati, S. (2016). Prevention and control policy analysis for energy-related regional pollution management in China. *Applied energy*, *166*, 292–300.
56. Liu, P., Zhou, Y., Zhou, D. K., & Xue, L. (2017). Energy Performance Contract models for the diffusion of green-manufacturing technologies in China: A stakeholder analysis from SMEs' perspective. *Energy Policy*, *106*, 59–67.
57. Lober, D. J. (1998). Pollution prevention as corporate entrepreneurship. *Journal of organizational change management*, *11*(1), 26–37.
58. Mamba, G., & Mishra, A. K. (2016). Graphitic carbon nitride (g-C3N4) nanocomposites: a new and exciting generation of visible light driven photocatalysts for environmental pollution remediation. *Applied Catalysis B: Environmental*, *198*, 347–377.

59. Maruthi, G. D., & Rashmi, R. (2015). Green Manufacturing: It's Tools and Techniques that can be implemented in Manufacturing Sectors. *Materials Today: Proceedings*, *2*(4–5), 3350–3355.
60. McGrath, D., Zhang, C., & Carton, O. T. (2004). Geostatistical analyses and hazard assessment on soil lead in Silvermines area, Ireland. *Environmental Pollution*, *127*(2), 239–248.
61. McNamara, M., Bohle, P., & Quinlan, M. (2011). Precarious employment, working hours, work-life conflict and health in hotel work. *Applied Ergonomics*, *42*(2), 225–232.
62. Merklein, M., Junker, D., Schaub, A., & Neubauer, F. (2016). Hybrid additive manufacturing technologies–an analysis regarding potentials and applications. *Physics Procedia*, *83*, 549–559.
63. Mittal, V. K., & Sangwan, K. S. (2014). Prioritizing barriers to green manufacturing: environmental, social and economic perspectives. *Procedia Cirp*, *17*, 559–564.
64. Montgomery, W. D. (1972). Markets in licenses and efficient pollution control programs. *Journal of Economic Theory*, *5*(3), 395–418.
65. Morrow, W. R., Qi, H., Kim, I., Mazumder, J., & Skerlos, S. J. (2007). Environmental aspects of laser-based and conventional tool and die manufacturing. *Journal of Cleaner Production*, *15*(10), 932–943.
66. Mueller, B. (2012). Additive manufacturing technologies–Rapid prototyping to direct digital manufacturing. *Assembly Automation*, *32*(2), 1–10.
67. Murugesan, S. (2008). Harnessing green IT: Principles and practices. *IT professional*, *10*(1), 24–33.
68. Nau, B., Roderburg, A., & Klocke, F. (2011). Ramp-up of hybrid manufacturing technologies. *CIRP Journal of Manufacturing Science and Technology*, *4*(3), 313–316.
69. Paris, H., Mokhtarian, H., Coatanéa, E., Museau, M., & Ituarte, I. F. (2016). Comparative environmental impacts of additive and subtractive manufacturing technologies. *CIRP Annals*, *65*(1), 29–32.
70. Paul, I. D., Bhole, G. P., & Chaudhari, J. R. (2014). A review on green manufacturing: it's important, methodology and its application. *Procedia Materials Science*, *6*, 1644–1649.
71. Petraru, M., & Gavrilescu, M. (2010). Pollution prevention, a key to economic and environmental sustainability. *Environmental Engineering and Management Journal*, *9*(4), 597–614.

72. Quinlan, M., Mayhew, C., & Bohle, P. (2001). The global expansion of precarious employment, work disorganization, and consequences for occupational health: a review of recent research. *International journal of health services*, *31*(2), 335–414.

73. Ramanathan, R. (2005). Estimating energy consumption of transport modes in India using DEA and application to energy and environmental policy. *Journal of the Operational Research Society*, *56*(6), 732–737.

74. Rew, A. (2007). Phytoremediation: an environmentally sound technology for pollution prevention, control and remediation in developing countries. *Educational Research and Reviews*, *2*(7), 151–156.

75. Rusinko, C. (2007). Green manufacturing: an evaluation of environmentally sustainable manufacturing practices and their impact on competitive outcomes. *IEEE Transactions on Engineering Management*, *54*(3), 445–454.

76. Salonitis, K., & Al Zarban, S. (2015). Redesign optimization for manufacturing using additive layer techniques. *Procedia CIRP*, *36*, 193–198.

77. Sangwan, K. S. (2006). Performance value analysis for justification of green manufacturing systems. *Journal of Advanced Manufacturing Systems*, *5*(01), 59–73.

78. Sarkis, J., & Cordeiro, J. J. (2001). An empirical evaluation of environmental efficiencies and firm performance: pollution prevention versus end-of-pipe practice. *European Journal of Operational Research*, *135*(1), 102–113.

79. Şchiopu, A. M., Apostol, I., Hodoreanu, M., & Gavrilescu, M. (2007). Solid Waste in Romania: Management, Treatment and Pollution Prevention Practices. *Environmental Engineering & Management Journal (EEMJ)*, *6*(5), 8–17.

80. Segerson, K. (1988). Uncertainty and incentives for nonpoint pollution control. *Journal of environmental economics and management*, *15*(1), 87–98.

81. Sezen, B., & Cankaya, S. Y. (2013). Effects of green manufacturing and eco-innovation on sustainability performance. *Procedia-Social and Behavioral Sciences*, *99*, 154–163.

82. Shen, T. T. (1995). Industrial pollution prevention. In *Industrial Pollution Prevention* (Vol. 1, pp. 15–35). Springer, Berlin, Heidelberg.

83. Shrivastava, P. (1995). Environmental technologies and competitive advantage. *Strategic Management Journal*, *16*(S1), 183–200.
84. Sirés, I., & Brillas, E. (2012). Remediation of water pollution caused by pharmaceutical residues based on electrochemical separation and degradation technologies: a review. *Environment International*, *40*, 212–229.
85. Sturgill, J. A., Swartzbaugh, J. T., & Randall, P. M. (2000). Pollution prevention in the semiconductor industry through recovery and recycling of gallium and arsenic from GaAs polishing wastes. *Clean Products and Processes*, *2*(1), 18–27.
86. Sullivan, R. M., & Floyd, D. W. (1991). Pollution prevention planning: An analysis of statutory structure for successful policy implementation. *Environ. Profess.*, *13*(4), 293–304.
87. Tsai, W. H., Chen, H. C., Leu, J. D., Chang, Y. C., & Lin, T. W. (2013). A product-mix decision model using green manufacturing technologies under activity-based costing. *Journal of Cleaner Production*, *57*, 178–187.
88. Underhill, E., & Quinlan, M. (2011). How precarious employment affects health and safety at work: the case of temporary agency workers. *Relations Industrielles/Industrial Relations*, *66*(3), 397–421.
89. Valls, M., & De Lorenzo, V. (2002). Exploiting the genetic and biochemical capacities of bacteria for the remediation of heavy metal pollution. *FEMS Microbiology Reviews*, *26*(4), 327–338.
90. Vaseashta, A., Vaclavikova, M., Vaseashta, S., Gallios, G., Roy, P., & Pummakarnchana, O. (2007). Nanostructures in environmental pollution detection, monitoring, and remediation. *Science and Technology of Advanced Materials*, *8*(1–2), 47.
91. Velayudham, A. (2007). Modern manufacturing processes: a review. *Journal on Design and Manufacturing Technologies*, *1*(1), 31.
92. Von Wandruszka, R. (2000). Humic acids: Their detergent qualities and potential uses in pollution remediation. *Geochemical Transactions*, *1*(1), 10.
93. Watanabe, C. (1999). Systems option for sustainable development—effect and limit of the Ministry of International Trade and Industry's efforts to substitute technology for energy. *Research Policy*, *28*(7), 719–749.

94. White, A. L., & Becker, M. (1992). Total cost assessment: catalyzing corporate self-interest in pollution prevention. *New Solutions: A Journal of Environmental and Occupational Health Policy*, *2*(3), 34–39.
95. Wu, H., Lai, C., Zeng, G., Liang, J., Chen, J., Xu, J., ... & Lu, L. (2017). The interactions of composting and biochar and their implications for soil amendment and pollution remediation: a review. *Critical Reviews in Biotechnology*, *37*(6), 754–764.
96. Xiaohui, L. F. Z. H. C. (1999). A Decision Making Framework Model of Green Manufacturing [J]. *Chinese Journal of Mechanical Engineering, 5, 10–21.*
97. Xun, S. (2017). The Mineral Industry of Iceland. *Minerals Yearbook: Area Reports: International Review 2013 Europe and Central Eurasia, 3, 1–12.*
98. Yongming, L. (2009). Trends in soil environmental pollution and the prevention-controlling-remediation strategies in China [J]. *Environmental Pollution & Control, 12, 1–10.*
99. Youngblood, D. J., Dvorak, B. I., Woldt, W. E., Hawkey, S. A., & Hygnstrom, J. R. (2008). Quantifying and comparing a P2 program's benefits: pollution prevention technical assistance in Nebraska. *Journal of Cleaner Production*, *16*(6), 761–770.
100. Zhou, Y., Pan, M., Zhou, D. K., & Xue, L. (2018). Stakeholder risk and trust perceptions in the diffusion of green manufacturing technologies: Evidence from China. *The Journal of Environment & Development*, *27*(1), 46–73.
101. Zhou, Y., Xu, G., Minshall, T., & Liu, P. (2015). How do public demonstration projects promote green-manufacturing technologies? A case study from China. *Sustainable Development*, *23*(4), 217–231.

CHAPTER 4

Prevention and Management of Environmental Pollution

CONTENTS

4.1. INTRODUCTION

Throughout the 20th century, the application and development of technologies and methods to lower the impacts of pollution have been prompted due to the increasing awareness of public health and environmental impacts because of the human activities (as discussed in the chapter related to Environmental Hazards). In this regard, several measures including regulatory and policymaking have been adopted by governments to lower the harmful impacts on the environment and to ensure that the quality standards of the environment are achieved (Harada, 1995; Samanta et al., 2002).

The basic goal of this chapter is to deliver an approach to prevent and control environmental pollution through the application of different methods. As we go along, we will introduce the main principles as a guide for the elimination of harmful effects on the air, water, and land quality; will consider the technological shift from control to the prevention of pollution; and will examine the limitations faced in generating solutions for separate environmental media. The mere removal of solid metal contaminants from air to make it pure and transferring these pollutants to land without any proper waste disposal system is not a solution to this problem. A combination of different solution media is needed for this purpose (Davis, 2002; Sures, 2004; Ren et al., 2011).

4.2. THE CONTAMINATION CONTROL APPROACH

The fast industrial development has not only created sites of contamination for air, water, and land resources with harmful pollutants and other materials but has also threatened ecosystems and humans with definitive health risks. The quality of global, regional, as well as local ecosystems have been intensely affected by the more intensive and extensive use of energy and materials (Hakanson, 1980; Tietenberg, 1995; 2002).

Before the intensive awareness of restricting environmental impacts of pollution, environmental management occurred, as a result, to avoid any local issue from a short period approach by disposing of waste materials beyond limitation. As the damage became unacceptable in certain areas, the requirement of a solution for this problem rose by exception. With the growth of industrial activity, grew the understanding of long-term effects and the main approach to ecological management came out to be the paradigm of controlling pollution (Milford et al., 1989; Long, 1992).

For this approach, the basis was formed by two precise concepts (Barbera & McConnell, 1986; Wong & Somes, 1995):

1. The concept of assimilative capacity, in which the assertion is made that only a certain level of environmental emissions is allowed which do not cause any harmful human health and environmental effects.
2. The concept of control principle, in which the assumption is made that by controlling the rate, time, and manner through which the disposal is made into the environment, the environmental damage can be prevented.

In the contamination control approach, the environment has been tried to be protected by relying on separating pollutants from the environment and using scrubbers and end-of-pipe filters (Lübbe-Wolff, 2001). The focus of these solutions has been tented to be on limitation of emissions or media-specific quality objectives and has primarily been directed at the source of discharge where the pollutants enter the environment let it be land, air, or water.

4.3. COMPREHENSIVE WASTE-MANAGEMENT

In this perspective of contamination control, waste is considered as an undesirable excessive product obtained as a result of the production process which needs to be disposed of carefully so that it does not affect the land, air, or water beyond and acceptable limit. An overview of the problems that must be considered in waste-management has been provided by Lucien Maystre in which a conceptual link has been provided to the progressively significant roles of pollution prevention and recycling (Penetrante, 1993; Pattrick et al., 2004).

To ensure the protection of the environment, several standards for the adequate practices for handling, collection, and disposal have been established by the governments, as a response to the extensive contamination of environment-related to unrestricted waste-management. The disposal of waste through incineration, sanitary landfills, and hazardous-waste treatment has gained specific attention to make sure that it is environmentally safe (Junhua et al., 2005; An et al., 2011).

Increasing attention has been paid to recycling and waste minimization to prevent the environmental impact and costs related to waste disposal and to promote reuse of existing limited resources. A summarized report has been provided by Niels Hahn and Poul Lauridsen in which the issues related to using recycling as a priority for waste-management strategy have been addressed considering its potential exposure implications to the worker (Hammer, 1999; Popp, 2006).

4.4. SHIFTING IMPORTANCE TO POLLUTION PREVENTION

Controlling pollution through end-of-pipe filtration can act as a risk of transferring pollutants from one source to another where it might either become a cause of equally severe environmental harm or even become an indirect pollution source for the same medium. Although this process is not as expensive as the treatment of the waste, it might become a financial risk by affecting the cost of the manufacturing process without adding any worth (Choe & Fraser, 1999; Minghua et al., 2009).

Although the approach of controlling pollution has been effective in achieving potential success in improving short-term local problems of pollution, it has failed to address the combined problems caused by pollution at a regional or global level, i.e., acid rain or ozone depletion, respectively (Kanagaraj et al., 2015; Zaman, 2015; 2016).

The reduction of pollution to the lowest possible level to enhance the quality of life is the basic objective of program-oriented health by controlling environmental pollution. All pollution aspects (i.e., land, air, and water, etc.) are covered by policies and programs of environmental contamination control, whose priorities and implications differ from land to land, and coordination among areas, i.e., city planning, industrial development, transportation policies, and water resources development is also involved in them (Eighmy et al., 1995; Hartlén, 1996).

The North-American Great Lakes is a case study that has been provided by Thomas Tseng, Victor Shantora and Ian Smith as an example of the environmental impacts of pollution caused by different media subjected to several stresses. A particular analysis of the restricted effectiveness of the contamination control model, in treating the pollutants that are emitted to the environment, has been made. Illustrations of the consequences of the practices that are related to control as well as prevention have been given by aiming at the method being followed in one land and its international repercussions (Hostovsky, 2006).

Along with reducing harmful impacts of industrial processes, to promote the affordability of industries, interests and being developed in taking preventive measures in the industrial process designs since the ecological contamination control technologies are expensive and non-feasible. Among many advantages of these preventive measures, reduction in toxic use and clean technologies are potential benefits in removing health risks for the worker (Boyle, 2000).

David Bennett gives a review of why contamination avoidance is developing as a favored methodology and how it identifies with other ecological administration techniques. This methodology is vital to actualizing the move to maintainable advancement which has been generally embraced since the arrival of the United Nations Commission on Trade and Development in 1987 and repeated at the Rio United Nations Conference on Environment and Development (UNCED) Conference in 1992.

The contamination removal approach centers legitimately around the utilization of procedures, materials, practices, and vitality that keep away from or limit the formation of wastes and toxins and at source, and not on "add-on" reduction measures. While corporate responsibility assumes a basic job in the choice to seek after contamination removal (check Bringer and Zoesel in Environmental strategy), Bennett causes to notice the cultural advantages in diminishing dangers to a biological system and human well-being and the strength of laborers specifically. He recognizes rules that can be conveniently applied in surveying possibilities for seeking after this methodology (Groot et al., 2014).

4.5. AIR CONTAMINATION MANAGEMENT

Air contamination management focuses on the end, or decrease to adequate levels, of suspended particulate issue, airborne vaporous toxins, and physical and, somewhat, natural living species whose presence in the air can cause unfriendly impacts on human wellbeing (e.g., disturbance, increment of occurrence or commonness of respiratory maladies, grimness, disease, overabundance mortality) or welfare (e.g., tangible impacts, decrease of visibility), injurious consequences for creature or vegetation, harm to materials of financial incentive to society and harm to the earth (e.g., climatic changes). The genuine perils related to radioactive toxins, just as the exceptional methodology required for their control and transfer, likewise merit cautious consideration (Shen, 1995; Bailey & Solomon, 2004).

The significance of productive administration of indoor and outdoor air contamination cannot be overstressed. Except if there is satisfactory control, the duplication of contamination sources in the advanced world may prompt hopeless harm to nature and humankind (Cooper et al., 1997; Li & Shue, 2004).

The target of this article is to give a general diagram of the potential ways to deal with the administration of surrounding air contamination from engine vehicle and mechanical sources. Nonetheless, it is to be accentuated from the earliest starting point that indoor air contamination (specifically, in developing nations) may assume a significantly bigger role than open-air contamination because of the perception that indoor air toxin concentrations are regularly considerably higher than open-air concentrations (McHugh et al., 1997; Gundersen et al., 2006).

Past contemplation of discharges from fixed or versatile sources, air contamination management includes thought of extra factors (e.g., meteorology and geography, and government and community investment, among numerous others) which must all be coordinated into a detailed program (Mage et al., 1996; Guttikunda et al., 2014). For instance, atmospheric conditions can extraordinarily influence the ground-level fixations coming about because of a similar toxin emanation. Air contamination sources might be dissipated over a network or a district and their impacts might be felt by, or their containing may include more than one organization. Moreover, air contamination does not regard any limits, and outflows from one locale may incite impacts in another area by long-separation transport (Sarkis & Cordeiro, 2001). Air contamination management, in this manner, needs a multidisciplinary method just as a joint exertion by private and administrative entities (Yang et al., 2008; Gilliland, 2009).

4.6. SOURCES OF AIR POLLUTION

The sources of human-made air contamination (or outflow sources) are of fundamentally two sorts (Freeman et al., 1992; Lober, 1998):

1. Stationary, which can be further divided into territory sources, for example, agrarian creation, quarrying and mining, mechanical, point and region sources, for example, assembling of synthetic compounds, non-metallic mineral items, fundamental metal corporations, power production and social sources (e.g., warming of homes and structures, civil waste and sewage muck incinerators,

chimneys, cooking places, clothing services and cleaning plants) (Gee & Sollars, 1998; Kowalczyk et al., 1978; Foley et al., 2010).

2. Mobile, involving any type of combustion-engine transports (e.g., light-duty gas controlled autos, light- and heavy-duty vehicles powered by diesel, bikes, air-vehicles, incorporating line sources with discharges of fumes and particulates from transport movement) (Pönkä, 1991; Barnett et al., 2006; Guo et al., 2013).

Additionally, natural contamination sources are also there (e.g., volcanoes, eroded areas, large amounts of pollen released by certain plants, sources of spores, bacteria, and viruses) (De Bortoli et al., 1986; Cabezas et al., 1999).

4.7. TYPES OF AIR CONTAMINANTS

Air contaminants are generally grouped into suspended particulates (fumes, dust, fogs, smokes), vaporous poisons (gases and fumes) and odors. A few instances of common toxins are given below:

Suspended particulate matter (SPM, PM-10) incorporates diesel fumes, coal fly-debris, mineral particles (e.g., coal, limestone, asbestos, and cement), metal particles and vapor (e.g., zinc, iron, copper, lead) and corrosive fogs (e.g., sulfuric acid), pesticide mists, fluorides, paint pigments, oil and carbon black smoke. Suspended particulate contaminants, other than their impacts of inciting respiratory sicknesses, malignant growths, erosion, obliteration of vegetation, etc., can likewise comprise an aggravation (e.g., nuisance of earth), meddle with sun-light (e.g., formation of haze and smog because of light dispersing) and go about as synergist surfaces for reaction of adsorbed synthetic substances (Freitas et al., 1997; Chuang et al., 2007).

Gaseous contaminants contain sulfur compounds (e.g., sulfur dioxide and trioxide), nitrogen compounds (e.g., nitric oxides and dioxides, ammonia), carbon monoxide, carbon-based compounds (e.g., hydrocarbons, polycyclic aromatic hydrocarbons (PAH), volatile organic compounds (VOC), aldehydes, halogen compounds and derivatives (e.g., HCl and HF), hydrogen sulfide, carbon disulfide and mercaptans (odors).

Secondary contaminants may be generated by reactions of chemical, thermal, or photochemical nature. For instance, sulfur trioxide can be formed by oxidation of sulfur dioxide by thermal activity which forms sulfuric acid mist when dissolved in water (catalyzed by iron and manganese oxides). The reaction of photochemical nature between reactive hydrocarbons and nitrogen oxides can produce formaldehyde, ozone (O_3), and peroxyacetyl nitrate (PAN); bis-chloromethyl ether can be formed by reactions between formaldehyde and HCl. While a few smells are known to be brought about by explicit chemical operators, for example, carbon disulfide (CS_2), hydrogen sulfide (HS), and mercaptans (R-SH or R1-S-R2), others are hard to characterize synthetically (Tronci et al., 2002; Lee and Wang, 2004). Instances of the primary toxins related to some mechanical air contamination sources are displayed in Table 4.1.

Table 4.1: Common Atmospheric Contaminants and Their Sources

Category	Emitted contaminants	Source
Agriculture	COC, CO, SPM	Open combustion
Mining and extraction	VOC, NO_x, SO_2, PM	Coal mining
	SO_2	Natural gas and crude petroleum gas production
	PB, SPM	Non-ferrous ore extraction
	SPM	Stone mining
Manufacturing	H_2S, VOC, CO, SPM	Tobacco, food, and beverages
	VOC, SPM	Leather and textiles industries
	VOC, SPM	Wood products
	R-SH, CO, SO_2, SPM, VOC, H_2S	Printing, paper products
Chemical syntheses	VOC, CO, SPM, SO_2	Phthalic anhydride
	Cl_2	Chlor-alkali
	HCl	Hydrochloric acid
	HF, SiF_4	Hydrofluoric acid
	SO_2, SO_3	Sulfuric acid
	NO_x	Nitric acid
	SPM, F_2	Phosphoric acid
	SPM, Pb	Pigments and lead oxide
	NO_x, CO, SPM, SO_2, VOC, NH_3	Ammonia
	SPM, NH_3	Sodium carbonate

	SPM	Calcium carbide
	NO_x, SPM, CO, VOC	Adipic acid
	Pb	Alkyl lead
	CO, VOC	Tetraphthalic acid and Maleic anhydride
	SPM, NH_3	Pesticide production and fertilizer
	NH_3, SPM, HNO_3	Ammonium nitrate
	VOC	Ammonium sulfate
	VOC, H_2S, SPM, CS_2	Plastic materials, synthetic resins, fibers
	SPM, VOC	Varnishes, paints, lacquers
	SPM	Soap
	NO_x, CO, SPM, SO_2, VOC, H_2S	Printing ink and carbon black
	NO_x, SO_3, SPM, SO_2, HNO_3	Trinitrotoluene
Petroleum refineries	NO_x, CO, SPM, SO_2, VOC	Miscellaneous products (coal and petroleum)
Manufacturing of mineral products (Non-metallic)	NO_x, CO, SPM, SO_2, VOC, F	Glass products
	NO_x, CO, SPM, SO_2, VOC, F_2	Structural clay goods
	SO_2, SPM, NO_x, CO	Lime, cement, and plaster
Basic metal manufacturing	NO_x, CO, SPM, SO_2, VOC, Pb	Steel and Iron
	SO_2, SPM, F, Pb	Non-ferrous industries
Power generation	NO_x, SPM, SO_2, CO, SO_3, VOC, Pb	Electricity, steam, and gas
Retail and wholesale trade	VOC	Filling operations, fuel storage
Transport	NO_x, CO, SPM, SO_2, VOC, Pb	
Community services	NO_x, CO, SPM, SO_2, VOC, Pb	Municipal incinerators

4.8. CLEAN AIR IMPLEMENTATION PLANS

Air quality administration focuses on the safeguarding of ecological quality by endorsing the endured level of contamination, leaving it to the nearby specialists and polluters to devise and execute activities to guarantee that this level of contamination will not be surpassed (Vinzents and Laursen,

1993; Pan et al., 2010). A case of enactment inside this methodology is the selection of encompassing air quality benchmarks based, all the time, on air quality rules for various contaminants; these are acknowledged greatest degrees of contaminants (or markers) in the objective zone (e.g., at ground level at a predefined point in a network) and can be either essential or optional measures (Currie, 1979; Yuhnke, 1991; Waltner, 1995). Essential measures are the most extreme levels reliable with a satisfactory security edge and with the safeguarding of general well-being, and must be followed inside a particular time limit; auxiliary benchmarks are those determined to be vital for assurance against known or foreseen antagonistic impacts other than well-being dangers (fundamentally on vegetation) and must be agreed "inside a sensible time." Air quality principles are long-, medium-, or short-term esteems substantial for 24 hours of the day, 7 days out of each week, and for month to month, regular or yearly introduction of every living subject (counting delicate subgroups, for example, kids, the old and the ill) as well as non-living objects; this is as opposed to most extreme passable levels for word related presentation, which is for an incomplete week by week exposure (e.g., 8 hours out of each day, 5 days out of each seven days) of grown-up and healthy workers (Chauhan et al., 2003; Curtis et al., 2006).

Conventional steps in air quality administration are control actions at the source, for instance, the compulsion of the utilization of synergistic converter systems in vehicles or of emanation guidelines in incinerators, planning of land-use and shut-down of processing plants or decrease of traffic during negative climate conditions (Chernow, 1974; Dwyer, 1995). The best air quality administration focuses on that the air contamination discharges ought to be kept to the lowest; this is fundamentally characterized through outflow principles for single points of emission of air contamination and could be accomplished for mechanical sources, for instance, through closed-systems and high-proficiency receivers. A discharge standard is a breaking point on the sum or grouping of contamination transmitted from a source (Chan et al., 2011; Vanos et al., 2013). This kind of enactment requires a choice, for every industry, on the best methods for controlling its emanations (i.e., fixing discharge gauges).

The essential point of air contamination administration is to determine a clean air execution plan (or air contamination reduction plan) which comprises of the below-given components (Szyszkowicz et al., 2009; Lei et al., 2011):

1. Description of the territory regarding geology, climate study and

socio-economy discharges stock.

2. Comparison with outflow measures.
3. Air contamination concentrations stock.
4. Simulated air contamination concentrations.
5. Comparison with air quality models.
6. Inventory of impacts on general well-being and the earth.
7. Causal investigation.
8. Control actions.
9. Cost of general well-being and ecological impacts.
10. Cost of control actions.
11. Cost-advantage investigation (expenses of control versus Expenses of efforts).
12. Transportation and planning of land-use.
13. Enforcement arrangement; asset responsibility.
14. Future projections on the populace, traffic, enterprises, and fuel utilization systems for development.

A portion of these issues will be depicted below.

4.8.1. Emanations Inventory; Correlation with Emanation Standards

The emanations inventory is a most-complete list of sources in a given region and their discharges, assessed as precisely as conceivable from all producing point, line and territory (diffuse) sources. At the point when these discharges are contrasted and outflow guidelines set for a specific source, the first indications on conceivable control measures are given if emanation models are not agreed to. The emanations inventory additionally serves to survey a need rundown of significant sources as per the measure of contaminants discharged, and shows the overall impact of various sources, for instance, traffic when contrasted with mechanical or private sources. The discharges inventory likewise permits a gauge of air contamination concentrations for those toxins for which surrounding concentration estimations are troublesome or too costly to even think about performing (Reitze Jr, 2004).

4.8.2. Air Contamination Concentrations Inventory; Correlation with Standards of Air Quality

The air contamination concentration inventory encapsulates the outcomes of the observation of ambient air contaminants in terms of yearly medians, percentiles, and trends of these amounts. Following are the compounds measured to be included for such an inventory (Mandelker & Rothschild, 1973):

1. Nitrogen oxides
2. Sulfur dioxide
3. Carbon monoxide
4. Suspended particulates
5. Heavy metals (Ni, Cu, Pb, Cd, Fe, Be, As)
6. Ozone
7. Polycyclic aromatic hydrocarbons: benzo (e) pyrene, benzo (a) pyrene, dibenzo (a, h) anthracene, benzo(a)anthracene, coronene, benzo (g, h, i) perylene
8. Volatile organic compounds: 3-methyl-hexane, *n*-heptane, toluene, *n*-hexane, benzene, *n*-nonane, isopropylbenzene, octane, ethyl-benzene xylene (*o*-, *m*-, *p*-), 1,2,4-/1,3,5-trimethyl benzene, trichloromethane, propyl benzene, *n*-2-/3-/4-ethyl toluene, tetrachloromethane, 1,1,1-trichloroethane, tri-/tetrachloroethene.

Correlation of air contaminant concentration with air quality rules or standards, on the off chance that they exist, shows issue regions for which a causal investigation must be acted to discover which sources are liable for the resistance. Scattering modeling must be utilized in playing out this causal investigation (see "Air contamination: Modeling of Air Contaminant Scattering"). Gadgets and systems utilized in the present surrounding air contamination observation are depicted in "Air quality monitoring."

4.8.3. Simulated Air Contaminant Concentrations; Correlation with Standards of Air Quality

Beginning from the discharges inventory, with its large number of materials which all cannot be checked in the surrounding air for financial reasons, utilization of scattering modeling can assess the concentrations of progressively "exotic" compounds. Utilizing fitting meteorology parameters in a reasonable scattering model, yearly midpoints and percentiles can be evaluated and contrasted with air quality norms or rules, in the event that they exist (Hahn, 1988).

4.8.4. Inventory of Impacts on General Well-being and the Atmosphere; Causal Investigation

Another significant wellspring of data is the impacts inventory (Ministerium für Umwelt 1993), which comprises of consequences of epidemiological investigations in the given territory and impacts of air contamination seen in material and biological receptors such as, plants, creatures, and structural metals and stones. Observed impacts ascribed to air contamination must be causally examined as for the element responsible for a specific impact, for instance, expanded predominance of ceaseless bronchitis in a contaminated region. In the event that the material or materials have been contained in a causal investigation (compound-causal examination), a subsequent investigation must be performed to discover the dependable sources (source-causal investigation).

4.8.5. Control Actions; Expenses of Control Actions

Control actions for industrial plants incorporate sufficient, well-planned, well-introduced, productively worked and kept up air cleaning gadgets, additionally called separators or gatherers. A separator or gatherer can be characterized as a "mechanical assembly for isolating any at least one of the accompanying from a vaporous medium in which they are mixed or suspended: strong particulates (filter and residue separators), fluid particles (filters and drop separator) and gases (gas cleaner)." The fundamental kinds of air contamination control hardware (talked about further in "Air contamination control") are the accompanying:

1. For particulates: inertial isolators (e.g., violent winds); texture filters (bag-houses); electrostatic precipitators; wet gatherers (scrubbers).
2. For vaporous contaminants: adsorption units (e.g., adsorption beds); wet gatherers (scrubbers); max-engine propulsion, which can be direct-fired (thermal burning) or synergist (reactant ignition).

Wet collectors (scrubbers) can be used to collect, at the same time, gaseous pollutants and particulate matter. Also, certain types of combustion devices can burn combustible gases and vapors as well as certain combustible aerosols. Depending on the type of effluent, one or a combination of more than one collector can be used (Beder, 1996; Amann et al., 2011).

By analogy with adequate practices at the workplace, good practices at the community level can contribute to air pollution control; for example, changes in the use of motor vehicles (more collective transportation, small cars and so on) and control of heating facilities (better insulation of buildings to require less heating, better fuels and so on).

Control actions in vehicle outflows are satisfactory and effective required review and upkeep programs which are authorized for the current vehicle armada, projects of implementation of the utilization of exhaust systems in new autos, forceful substitution of sun oriented/battery-powered vehicles for fuel-controlled ones, guideline of street traffic, and transportation and land use planning ideas.

Engine vehicle discharges are constrained by controlling emanations per vehicle mile traveled (VMT) and by constraining VMT itself (Walsh 1992). Emanations per VMT can be diminished by controlling vehicle execution (equipment, upkeep), for both new and being used autos. Fuel creation of leaded gas might be constrained by diminishing lead or sulfur content, which additionally beneficially affects diminishing HC discharges from vehicles. Bringing down the degrees of sulfur in diesel fuel as a way to bring down diesel particulate emanation has the extra useful impact of expanding the potential for reactant control of diesel particulate and natural HC outflows (You & Xu, 2010).

Another significant administration device for diminishing vehicle evaporative and refueling outflows is the control of fuel volatility. Control of fuel instability can extraordinarily bring down vehicle evaporative HC outflows. The utilization of oxygenated additive substances in gas brings down HC and CO fumes insofar as fuel instability is not expanded (Seinfeld and Kyan, 1971; Krupnick and Portney, 1991).

Decrease of VMT is an extra method for controlling vehicle emanations by control techniques, for example,

1. Use of increasingly productive transportation modes.
2. Increasing the normal number of travelers per vehicle.
3. Spreading blocked pinnacle traffic loads.
4. Reducing the travel request.

While such methodologies advance fuel protection, they are not yet acknowledged by the all-inclusive community, and governments have not genuinely attempted to actualize them (Duda and Johnson, 1985; Xu et al.,

2000). All these mechanical and political answers for the engine vehicle issue aside from substitution of electrical autos are progressively balanced by development in the vehicle populace. The vehicle issue can be tackled just if the development issue is tended to in a proper manner (Schöpp et al., 1998; Martínez and Albiac, 2006).

4.8.6. Expenses of General Well-being and Environmental Impacts; Cost-Benefit Investigation

The assessment of the expenses of general well-being and ecological impacts is the most troublesome piece of a clean air execution plan, as it is hard to appraise the estimation of lifetime decrease of impairing ailments, emergency clinic affirmation rates and long periods of work lost. Be that as it may, this estimation and an examination with the expense of control actions is totally important so as to adjust the expenses of control actions versus the expenses of no such actions performed, regarding general well-being and natural impacts.

4.8.7. Land-Use Planning and Transportation

The contamination issue is personally associated with land-use and transportation, including problems, for example, network arranging, street configuration, traffic control, and mass transportation; to problems of demography, geography and economy; and to social concerns (Strauch et al., 2005; Barton, 2009). Generally, the quickly developing urban accumulations have extreme contamination issues because of poor land-use and transportation actions. Transportation arrangements for air contamination control incorporate transportation strategies, transportation controls, and mass travel and parkway clog costs. Transportation controls importantly affect the overall population as far as value, abusiveness and social and financial interruption, specifically, direct transportation controls, for example, engine vehicle imperatives, gas restrictions, and engine vehicle emanation decrease. Emanation decreases because of direct controls can be dependably evaluated and checked (Herala, 2003; Tiwari, 2003). Backhanded transportation controls, for example, decrease of vehicle miles went by progress of mass travel frameworks, traffic stream improvement guidelines, guidelines on parking garages, street and fuel charges, vehicle use consents and motivating forces for intentional methodologies are generally founded on past experimentation experience, and incorporate numerous vulnerabilities when attempting to build up a feasible transportation plan

(Affum et al., 2003). National activity plans bringing about aberrant transportation controls can influence transportation and land-use design as to roadways, parking areas, and shopping malls. Long haul getting ready for the transportation framework and the region impacted by it will forestall critical crumbling of air quality and furnish for consistency with air quality measures. Mass travel is reliably considered as a potential answer for urban air contamination issues. Determination of a mass travel framework to serve a territory and distinctive modular parts between expressway use and transport or rail administration will at last change land-use designs. There is an ideal part that will limit air contamination; nonetheless, this may not be adequate when non-ecological elements are considered (Banister, 1999).

The car has been known as the best generator of monetary externalities at any point known. A portion of these, for example, occupations and versatility, are sure, yet the negative ones, for example, air contamination, mishaps bringing about death and damage, property harm, clamor, time loss, and irritation, lead to the end that transportation is not a diminishing cost industry in urbanized zones. Thruway blockage costs are another externality; lost time and clog costs, nonetheless, are hard to decide. A genuine assessment of contending transportation modes, for example, mass transportation, cannot be acquired if travel costs for work trips do exclude clog costs.

Land-use anticipating air contamination control incorporates zoning codes and execution benchmarks, land-use controls, lodging and land advancement, and land-use design arrangements. Land-use zoning was the underlying endeavor to achieve the security of the individuals, their property and their financial chance. In any case, the universal idea of air contaminants required more than physical detachment of ventures and local locations to secure the person. Consequently, execution models dependent on aesthetics or subjective choices were brought into some zoning codes trying to evaluate criteria for distinguishing potential issues.

The confinements of the assimilative limit of the earth must be distinguished for long haul land-use design. At that point, land-use controls can be built up that will customize the limit impartially among wanted neighborhood exercises. Land-use controls incorporate grant frameworks for the audit of new stationary sources, zoning guidelines among mechanical and local locations, limitation by easement or acquisition of land, receptor area control, outflow thickness zoning, and emanation allotment guidelines.

Lodging arrangements planned for making home possession accessible to numerous who could otherwise not bear the cost of it (e.g., charge motivating forces and home loan strategies) invigorate urban spread and in a roundabout way demoralize higher-thickness private advancement. These approaches have now demonstrated to be ecologically heartbreaking, as no thought was given to the concurrent improvement of productive transportation frameworks to serve the necessities of the huge number of new networks being created. The exercise gained from this advancement is that projects affecting on nature ought to be composed, and thorough arranging embraced at the level where the issue happens and on a scale sufficiently huge to incorporate the whole framework.

Land-use design must be inspected at national, commonplace or state, local, and nearby levels to substantially guarantee long haul insurance of the environment. Legislative projects generally start with power plant siting, mineral extraction destinations, beachfront zoning and desert, mountain or other recreational improvements. As the variety of neighborhood governments in a given district can't satisfactorily manage territorial natural issues, provincial governments or organizations should facilitate land advancement and thickness designs by overseeing the spatial course of action and area of new development and use, and transportation offices. Land-use and transportation design must be interrelated with the requirement of guidelines to keep up the ideal air quality. In a perfect world, air contamination control ought to be gotten ready for by the equivalent provincial organization that lands-use design because of the covering externalities related with the two issues (Cavallin et al., 1994; Jay et al., 2007).

4.8.8. Resource management and Enforcement Plan

The clean air execution plan ought to consistently contain an authorization plan which demonstrates how the control actions can be upheld. This suggests likewise an asset responsibility which, as per a polluter pays standard, will state what the polluter needs to execute and how the administration will help the polluter in satisfying the dedication.

4.8.9. Future Projections

In the feeling of a preparatory arrangement, the clean air execution plan ought to likewise remember appraisals of the patterns for populace, traffic, businesses and fuel utilization so as to evaluate reactions to future issues. This will maintain a strategic distance from future worries by upholding estimates well ahead of time of envisioned issues.

4.8.10. Follow-up Strategies

A follow-up strategy of air quality administration comprises of strategies and policies on ways to execute future clean air operation strategies.

4.8.11. Role of Ecological Impact Analysis

Ecological impact analysis (EIA) is the way toward giving an itemized proclamation by the answerable agency on the natural effect of a proposed activity altogether influencing the nature of the human condition (Hollick, 1981). EIA is an instrument of counteraction focusing on the thought of the human condition at the beginning time of the advancement of a program or task. EIA is especially significant for nations that create extends in the structure of financial reorientation and rebuilding. EIA has become enactment in many developed nations and is currently progressively applied in developing nations and economies experiencing significant change (Morrison-Saunders et al., 2001; Cashmore, 2004).

EIA is integrative in the way of exhaustive natural planning and administration thinking about the connections between various ecological media. Then again, EIA incorporates the estimation of ecological results into the design procedure and in this way turns into an instrument of reasonable improvement. EIA additionally joins specialized and participative properties as it gathers, examinations and applies logical and specialized information with the thought of value control and quality confirmation, and stresses the significance of consultations preceding permitting techniques between ecological agencies and the population which could be influenced by specific undertakings (Alshuwaikhat, 2005; Ding, 2008). A clean air usage plan can be considered as a piece of the EIA methodology concerning the air.

REFERENCES

1. Affum, J. K., Brown, A. L., & Chan, Y. C. (2003). Integrating air pollution modelling with scenario testing in road transport planning: the TRAEMS approach. *Science of the Total Environment, 312*(1–3), 1–14.
2. Alshuwaikhat, H. M. (2005). Strategic environmental assessment can help solve environmental impact assessment failures in developing countries. *Environmental Impact Assessment Review, 25*(4), 307–317.
3. Amann, M., Bertok, I., Borken-Kleefeld, J., Cofala, J., Heyes, C., Höglund-Isaksson, L., ... & Sandler, R. (2011). Cost-effective control of air quality and greenhouse gases in Europe: Modeling and policy applications. *Environmental Modelling & Software, 26*(12), 1489–1501.
4. An, H. T. Q., Huu, T. P., Le Van, T., Cormier, J. M., & Khacef, A. (2011). Application of atmospheric non thermal plasma-catalysis hybrid system for air pollution control: Toluene removal. *Catalysis Today, 176*(1), 474–477.
5. Bailey, D., & Solomon, G. (2004). Pollution prevention at ports: clearing the air. *Environmental Impact Assessment Review, 24*(7–8), 749–774.
6. Banister, D. (1999). Planning more to travel less: Land use and transport. *Town Planning Review, 70*(3), 313.
7. Barbera, A. J., & McConnell, V. D. (1986). Effects of pollution control on industry productivity: a factor demand approach. *The Journal of Industrial Economics*, 1, 161–172.
8. Barnett, A. G., Williams, G. M., Schwartz, J., Best, T. L., Neller, A. H., Petroeschevsky, A. L., & Simpson, R. W. (2006). The effects of air pollution on hospitalizations for cardiovascular disease in elderly people in Australian and New Zealand cities. *Environmental Health Perspectives, 114*(7), 1018–1023.
9. Barton, H. (2009). Land use planning and health and well-being. *Land Use Policy, 26*, S115–S123.
10. Beder, S. (1996). Charging the earth: the promotion of price-based measures for pollution control. *Ecological Economics, 16*(1), 51–63.
11. Boyle, C. A. (2000). Solid waste management in New Zealand. *Waste Management, 20*(7), 517–526.

12. Cabezas, H., Bare, J. C., & Mallick, S. K. (1999). Pollution prevention with chemical process simulators: the generalized waste reduction (WAR) algorithm—full version. *Computers & Chemical Engineering*, *23*(4–5), 623–634.
13. Cashmore, M. (2004). The role of science in environmental impact assessment: process and procedure versus purpose in the development of theory. *Environmental Impact Assessment Review*, *24*(4), 403–426.
14. Cavallin, A., Marchetti, M., Panizza, M., & Soldati, M. (1994). The role of geomorphology in environmental impact assessment. *Geomorphology*, *9*(2), 143–153.
15. Chan, Y. C., Hawas, O., Hawker, D., Vowles, P., Cohen, D. D., Stelcer, E., … & Christensen, E. (2011). Using multiple type composition data and wind data in PMF analysis to apportion and locate sources of air pollutants. *Atmospheric Environment*, *45*(2), 439–449.
16. Chauhan, R. P., Kant, K., Sharma, S. K., & Chakarvarti, S. K. (2003). Measurement of alpha radioactive air pollutants in fly ash brick dwellings. *Radiation Measurements*, *36*(1–6), 533–536.
17. Chernow, E. (1974). Implementing the Clean Air Act in Los Angeles: The Duty to Achieve the Impossible. *Ecology LQ*, *4*, 537.
18. Choe, C., & Fraser, I. (1999). An economic analysis of household waste management. *Journal of Environmental Economics and Management*, *38*(2), 234–246.
19. Chuang, K. J., Chan, C. C., Su, T. C., Lee, C. T., & Tang, C. S. (2007). The effect of urban air pollution on inflammation, oxidative stress, coagulation, and autonomic dysfunction in young adults. *American Journal of Respiratory and Critical Care Medicine*, *176*(4), 370–376.
20. Cooper, W. W., Hemphill, H., Huang, Z., Li, S., Lelas, V., & Sullivan, D. W. (1997). Survey of mathematical programming models in air pollution management. *European Journal of Operational Research*, *96*(1), 1–35.
21. Currie, D. P. (1979). Relaxation of Implementation Plans Under the 1977 Clean Air Act Amendments. *Mich. L. Rev.*, *78*, 155.
22. Curtis, L., Rea, W., Smith-Willis, P., Fenyves, E., & Pan, Y. (2006). Adverse health effects of outdoor air pollutants. *Environment International*, *32*(6), 815–830.
23. Davis, D. L. (2002). *When Smoke Ran Like Water: Tales of Environmental Deception and the Battle Against Pollution* (Vol. 1, p.

45). New York: Basic Books.

24. De Bortoli, M. H. E. A. L. H. H. H., Knöppel, H., Pecchio, E., Peil, A., Rogora, L., Schauenburg, H., ... & Vissers, H. (1986). Concentrations of selected organic pollutants in indoor and outdoor air in northern Italy. *Environment International, 12*(1–4), 343–350.
25. Ding, G. K. (2008). Sustainable construction—The role of environmental assessment tools. *Journal of Environmental Management, 86*(3), 451–464.
26. Duda, A. M., & Johnson, R. J. (1985). Cost-effective targeting of agricultural nonpoint-source pollution controls. *Journal of Soil and Water Conservation, 40*(1), 108–111.
27. Dwyer, J. P. (1995). The Practice of Federalism Under the Clean Air Act. *Md. L. Rev., 54*, 1183.
28. Eighmy, T. T., Eusden, J. D., Krzanowski, J. E., Domingo, D. S., Staempfli, D., Martin, J. R., & Erickson, P. M. (1995). Comprehensive approach toward understanding element speciation and leaching behavior in municipal solid waste incineration electrostatic precipitator ash. *Environmental Science & Technology, 29*(3), 629–646.
29. Foley, K. M., Roselle, S. J., Appel, K. W., Bhave, P. V., Pleim, J. E., Otte, T. L., ... & Nolte, C. G. (2010). Incremental testing of the Community Multiscale Air Quality (CMAQ) modeling system version 4.7. *Geoscientific Model Development, 3*(1), 205–226.
30. Freeman, H., Harten, T., Springer, J., Randall, P., Curran, M. A., & Stone, K. (1992). Industrial pollution prevention! A critical review. *Journal of the Air & Waste Management Association, 42*(5), 618–656.
31. Freitas, M. C., Reis, M. A., Alves, L. C., Wolterbeek, H. T., Verburg, T., & Gouveia, M. A. (1997). Bio-monitoring of trace-element air pollution in Portugal: Qualitative survey. *Journal of Radioanalytical and Nuclear Chemistry, 217*(1), 21–30.
32. Gee, I. L., & Sollars, C. J. (1998). Ambient air levels of volatile organic compounds in Latin American and Asian cities. *Chemosphere, 36*(11), 2497–2506.
33. Gilliland, F. D. (2009). Outdoor air pollution, genetic susceptibility, and asthma management: opportunities for intervention to reduce the burden of asthma. *Pediatrics, 123*(Supplement 3), S168–S173.
34. Groot, J., Bing, X., Bos-Brouwers, H., & Bloemhof-Ruwaard, J. (2014). A comprehensive waste collection cost model applied to post-

consumer plastic packaging waste. *Resources, Conservation and Recycling*, *85*, 79–87.

35. Gundersen, P., Schmidt, I. K., & Raulund-Rasmussen, K. (2006). Leaching of nitrate from temperate forests effects of air pollution and forest management. *Environmental Reviews*, *14*(1), 1–57.
36. Guo, Y., Li, S., Tian, Z., Pan, X., Zhang, J., & Williams, G. (2013). The burden of air pollution on years of life lost in Beijing, China, 2004–08: retrospective regression analysis of daily deaths. *Bmj*, *347*, f7139.
37. Guttikunda, S. K., Goel, R., & Pant, P. (2014). Nature of air pollution, emission sources, and management in the Indian cities. *Atmospheric Environment*, *95*, 501–510.
38. Hahn, R. W. (1988). Innovative approaches for revising the Clean Air Act. *Nat. Resources J.*, *28*, 171.
39. Hakanson, L. (1980). An ecological risk index for aquatic pollution control. A sedimentological approach. *Water Research*, *14*(8), 975–1001.
40. Hammer, T. (1999). Application of plasma technology in environmental techniques. *Contributions to Plasma Physics*, *39*(5), 441–462.
41. Harada, M. (1995). Minamata disease: methylmercury poisoning in Japan caused by environmental pollution. *Critical Reviews in Toxicology*, *25*(1), 1–24.
42. Hartlén, J. (1996). Waste management in Sweden. *Waste Management*, *16*(5–6), 1–22.
43. Herala, N. (2003). Regulating traffic with land use planning. *Sustainable Development*, *11*(2), 91–102.
44. Hollick, M. (1981). Role of quantitative decision-making methods in environmental-impact assessment. *J. Environ. Manage.;(United States)*, *12*(1), 2–29.
45. Hostovsky, C. (2006). The paradox of the rational comprehensive model of planning: Tales from waste management planning in Ontario, Canada. *Journal of Planning Education and Research*, *25*(4), 382–395.
46. Jay, S., Jones, C., Slinn, P., & Wood, C. (2007). Environmental impact assessment: Retrospect and prospect. *Environmental impact assessment review*, *27*(4), 287–300.
47. Junhua, L., Jianjun, C., & Jiming, H. (2005). Recent progress of air pollution control technology. *Chemical Industry and Engineering Progress*, *24*(7), 703.

48. Kanagaraj, J., Senthilvelan, T., Panda, R. C., & Kavitha, S. (2015). Eco-friendly waste management strategies for greener environment towards sustainable development in leather industry: a comprehensive review. *Journal of Cleaner Production*, *89*, 1–17.
49. Kowalczyk, G. S., Choquette, C. E., & Gordon, G. E. (1978). Chemical element balances and identification of air pollution sources in Washington, DC. *Atmospheric Environment (1978)*, *12*(5), 1143–1153.
50. Krupnick, A. J., & Portney, P. R. (1991). Controlling urban air pollution: a benefit-cost assessment. *Science*, *252*(5005), 522–528.
51. Lee, S. C., & Wang, B. (2004). Characteristics of emissions of air pollutants from burning of incense in a large environmental chamber. *Atmospheric Environment*, *38*(7), 941–951.
52. Lei, Y., Zhang, Q., Nielsen, C., & He, K. (2011). An inventory of primary air pollutants and CO2 emissions from cement production in China, 1990–2020. *Atmospheric Environment*, *45*(1), 147–154.
53. Li, S. T., & Shue, L. Y. (2004). Data mining to aid policy making in air pollution management. *Expert Systems with Applications*, *27*(3), 331–340.
54. Lober, D. J. (1998). Pollution prevention as corporate entrepreneurship. *Journal of Organizational Change Management*, *11*(1), 26–37.
55. Long, N. V. (1992). Pollution control: A differential game approach. *Annals of Operations Research*, *37*(1), 283–296.
56. Lübbe-Wolff, G. (2001). Efficient environmental legislation-On different philosophies of pollution control in Europe. *Journal of Environmental Law*, *13*(1), 79–87.
57. Mage, D., Ozolins, G., Peterson, P., Webster, A., Orthofer, R., Vandeweerd, V., & Gwynne, M. (1996). Urban air pollution in megacities of the world. *Atmospheric Environment*, *30*(5), 681–686.
58. Mandelker, D. R., & Rothschild, S. B. (1973). The Role of Land-Use Controls in Combating Air Pollution Under the Clean Air Act of 1970. *Ecology lQ*, *3*, 235.
59. Martínez, Y., & Albiac, J. (2006). Nitrate pollution control under soil heterogeneity. *Land Use Policy*, *23*(4), 521–532.
60. McHugh, C. A., Carruthers, D. J., & Edmunds, H. A. (1997). ADMS–Urban: an air quality management system for traffic, domestic and industrial pollution. *International Journal of Environment and Pollution*, *8*(3–6), 666–674.

61. Milford, J. B., Russell, A. G., & McRae, G. J. (1989). A new approach to photochemical pollution control: Implications of spatial patterns in pollutant responses to reductions in nitrogen oxides and reactive organic gas emissions. *Environmental Science & Technology*, *23*(10), 1290–1301.
62. Minghua, Z., Xiumin, F., Rovetta, A., Qichang, H., Vicentini, F., Bingkai, L., ... & Yi, L. (2009). Municipal solid waste management in Pudong new area, China. *Waste Management*, *29*(3), 1227–1233.
63. Morrison-Saunders, A., Arts, J., Baker, J., & Caldwell, P. (2001). Roles and stakes in environmental impact assessment follow-up. *Impact Assessment and Project Appraisal*, *19*(4), 289–296.
64. Pan, G., Zhang, S., Feng, Y., Takahashi, K., Kagawa, J., Yu, L., ... & Pan, B. (2010). Air pollution and children's respiratory symptoms in six cities of Northern China. *Respiratory Medicine*, *104*(12), 1903–1911.
65. Pattrick, G., Van der Lingen, E., Corti, C. W., Holliday, R. J., & Thompson, D. T. (2004). The potential for use of gold in automotive pollution control technologies: a short review. *Topics in catalysis*, *30*(1–4), 273–279.
66. Penetrante, B. M. (1993, June). Pollution control applications of pulsed power technology. In *Ninth IEEE International Pulsed Power Conference* (Vol. 1, p. 1–22). IEEE.
67. Pönkä, A. (1991). Asthma and low level air pollution in Helsinki. *Archives of Environmental Health: An International Journal*, *46*(5), 262–270.
68. Popp, D. (2006). International innovation and diffusion of air pollution control technologies: the effects of NOX and SO2 regulation in the US, Japan, and Germany. *Journal of Environmental Economics and Management*, *51*(1), 46–71.
69. Reitze Jr, A. W. (2004). Air Quality Protection Using State Implementation Plans-Thirty-Seven Years of Increasing Complexity. *Vill. Envtl. LJ*, *15*, 209.
70. Ren, X., Chen, C., Nagatsu, M., & Wang, X. (2011). Carbon nanotubes as adsorbents in environmental pollution management: a review. *Chemical Engineering Journal*, *170*(2–3), 395–410.
71. Samanta, S. K., Singh, O. V., & Jain, R. K. (2002). Polycyclic aromatic hydrocarbons: environmental pollution and bioremediation. *TRENDS in Biotechnology*, *20*(6), 243–248.

72. Sarkis, J., & Cordeiro, J. J. (2001). An empirical evaluation of environmental efficiencies and firm performance: pollution prevention versus end-of-pipe practice. *European Journal of Operational Research, 135*(1), 102–113.
73. Schöpp, W., Amann, M., Cofala, J., Heyes, C., & Klimont, Z. (1998). Integrated assessment of European air pollution emission control strategies. *Environmental Modelling & Software*, *14*(1), 1–9.
74. Seinfeld, J. H., & Kyan, C. P. (1971). Determination of optimal air pollution control strategies. *Socio-Economic Planning Sciences*, *5*(3), 173–190.
75. Shen, T. T. (1995). Industrial pollution prevention. In *Industrial Pollution Prevention* (Vol.1, pp. 15–35). Springer, Berlin, Heidelberg.
76. Strauch, D., Moeckel, R., Wegener, M., Gräfe, J., Mühlhans, H., Rindsfüser, G., & Beckmann, K. J. (2005). Linking transport and land use planning: the microscopic dynamic simulation model ILUMASS. *Geodynamics*, 1, 295–311.
77. Sures, B. (2004). Environmental parasitology: relevancy of parasites in monitoring environmental pollution. *Trends in Parasitology*, *20*(4), 170–177.
78. Szyszkowicz, M., Stieb, D. M., & Rowe, B. H. (2009). Air pollution and daily ED visits for migraine and headache in Edmonton, Canada. *The American Journal of Emergency Medicine*, *27*(4), 391–396.
79. Tietenberg, T. (1995). Tradeable permits for pollution control when emission location matters: what have we learned?. *Environmental and Resource Economics*, *5*(2), 95–113.
80. Tietenberg, T. (2002). 19 Lessons from using transferable permits to control air pollution in the United States. *Handbook of Environmental and Resource Economics*, 275, 1–20.
81. Tiwari, G. (2003). Transport and land-use policies in Delhi. *Bulletin of the World Health Organization*, *81*, 444–450.
82. Tronci, S., Baratti, R., & Servida, A. (2002). Monitoring pollutant emissions in a 4.8 MW power plant through neural network. *Neurocomputing*, *43*(1–4), 3–15.
83. Vanos, J. K., Cakmak, S., Bristow, C., Brion, V., Tremblay, N., Martin, S. L., & Sheridan, S. S. (2013). Synoptic weather typing applied to air pollution mortality among the elderly in 10 Canadian cities. *Environmental Research*, *126*, 66–75.

84. Vinzents, P., & Laursen, B. (1993). A national cross-sectional study of the working environment in the Danish wood and furniture industry—air pollution and noise. *The Annals of Occupational Hygiene*, *37*(1), 25–34.

85. Waltner, A. C. (1995). Paradise Delayed--The Continuing Saga of the Los Angeles Basin Federal Clean Air Implementation Plan. *UCLA J. Envtl. L. & Pol'y*, *14*, 247.

86. Wong, T. H., & Somes, N. L. G. (1995). A stochastic approach to designing wetlands for stormwater pollution control. *Water Science and Technology*, *32*(1), 145.

87. Xu, X., Chen, C., Qi, H., He, R., You, C., & Xiang, G. (2000). Development of coal combustion pollution control for SO2 and NOx in China. *Fuel Processing Technology*, *62*(2–3), 153–160.

88. Yang, J., Yu, Q., & Gong, P. (2008). Quantifying air pollution removal by green roofs in Chicago. *Atmospheric Environment*, *42*(31), 7266–7273.

89. You, C. A., & Xu, X. C. (2010). Coal combustion and its pollution control in China. *Energy*, *35*(11), 4467–4472.

90. Yuhnke, R. E. (1991). The Amendments to Reform Transportation Planning in the Clean Air Act Amendments of 1990. *Tul. Envtl. LJ*, *5*, 239.

91. Zaman, A. U. (2015). A comprehensive review of the development of zero waste management: lessons learned and guidelines. *Journal of Cleaner Production*, *91*, 12–25.

92. Zaman, A. U. (2016). A comprehensive study of the environmental and economic benefits of resource recovery from global waste management systems. *Journal of Cleaner Production*, *124*, 41–50.

CHAPTER 5

Air Pollution Control

CONTENTS

5.1. INTRODUCTION

The aim of the manager of the air pollution control system is to confirm that unnecessary concentrations of air pollutants don't reach the susceptible target. Targets could comprise people, animals, plants and materials. In all circumstances, one should be anxious about the most sensitive of every one of these groups. Air pollutants could comprise gases, aerosols, vapors and, in some circumstances, bio-hazardous materials. The well-designed system will avert the target from receiving a dangerous concentration of the pollutant (Milford et al., 1989; Long, 1992). Most of the air pollution control systems include an amalgamation of various control techniques, normally an amalgamation of administrative controls and technological controls, and in complex or larger sources there might be more than one kind of technological control. Ideally, the choice of suitable controls will generally be made in the context of the issue to be solved (Hakanson, 1980; Tietenberg, 1995; 2002).

1. What is released, in what concentration?
2. Who are the targets? What is the most vulnerable target?
3. What are tolerable long-term exposure levels?
4. What are tolerable short-term exposure levels?
5. What amalgamation of controls should be chosen to confirm that the long-term and short-term exposure levels aren't exceeded?

Table 5.1: Steps in Choosing Pollution Controls

Step 1: Define emissions.	The first step is to conclude what will be emitted from the stack. All potentially injurious emissions should be listed. The second step is to assess how much of every material will be discharged. Without this basic information, the manager can't start to design the control program.
Step 2: Define target groups.	All vulnerable targets must be identified. This comprises people, plants, animals and materials. In every case, the most vulnerable member of every group should be identified. For instance, asthmatics near the plant that releases isocyanates.
Step 3: Determine tolerable exposure levels.	The tolerable level of exposure for the most subtle target group should be established. If the contaminant is a material that possesses cumulative effects, like carcinogen, then the long-term exposure levels must be set. If the contaminant possesses short-term effects, like a sensitizer or an irritant, then a short-term or possibly peak exposure level should be set.

Step 4: Select controls.	Step 1 recognizes the releases and Step 3 concludes the acceptable level. In this specific step, each contaminant is checked to confirm that it doesn't exceed the tolerable level. If it surpasses the tolerable level, extra controls should be added, and the levels of exposure checked again. This process carries on until all the exposures are below or ar the acceptable level. Dispersion modeling can be utilized to assess exposures for the new plants.

* When setting the levels of exposure in Step 3, it should be kept in mind that these levels of exposures are the total exposures, not only those from the plants. Once the tolerable level has been recognized, background levels and the contributions from some other plants should be subtracted to conclude the maximum amount that the plants can release without surpassing the tolerable exposure level. If this isn't done, and three plants are permitted to release at the maximum, then the groups of targets will be exposed to three times the tolerable level (Wong & Somes, 1995; Lübbe-Wolff, 2001).

Some materials like carcinogens don't have the verge below which no dangerous effects will occur. Thus, as long as some amount of the material is permitted to discharge to the environment, there will be a risk to the target populations. In this situation, a no-effect level can't be set. Rather, a tolerable level of risk should be established. Normally this is in the range of one adverse result in 100 thousand to 1000 thousand exposed persons (Barbera & McConnell, 1986).

Some authorities have done some work by setting the standards based on the maximum concentration of the pollutant that the susceptible target can accept. With this kind of standard, the manager doesn't need to carry out Steps 2 & 3, as the regulating agency has done this already. Under this system, the manager should develop only the uncontrolled discharge standards for every pollutant (Step 1), and then decide what controls are essential to meet the standards (Step 4).

5.2. MANAGING AIR POLLUTION

By having standards of air quality, the regulators can measure separate exposures and thus conclude whether someone is exposed to possibly harmful levels. It is supposed that the standards made under these conditions are quite low to safeguard the most vulnerable target group (Penetrante, 1993; Pattrick et al., 2004).

This isn't always a safe supposition. As exhibited in Table 5.2, there can be an extensive change in the common air quality standards. Air quality standards for SO_2 (sulfur dioxide) range from 30–140 µg/m, For the less commonly controlled materials, this change can be even larger, as displayed in Table 5.3, for benzene. This isn't surprising given that economics can play a role in setting the standards as the toxicology does. If the standard isn't set too low enough to safeguard vulnerable populations, no one is served well. Exposed populations have the feeling of false assurance, and can unintentionally be put at risk. The emitter might feel at first that they have profited from the lenient standard, but if impacts in the community need the company to reform their controls, or put in new controls, expenses could be very higher than doing it appropriately the first time (Junhua et al., 2005).

Table 5.2: Range of the Air Quality Standards for Controlled Air Pollutant (Sulfur Dioxide)

Countries and territories	**Long-term SO_2 air quality standards (µg/m^3)**
Australia	50
Finland	40
Canada	30
Hungary	70
Germany	140
Taiwan	133

Table 5.3: Range of the Air Quality Standards for Less Frequently Controlled Air Pollutant (Benzene)

City/State	**24-hour air quality standard for benzene (µg/m^3)**
Connecticut	53.4
Michigan	2.4
Massachusetts	1.2
Nevada	254
North Carolina	2.1
Philadelphia	1,327
New York	1,718
Virginia	300

The levels were generally standardized to an average time of almost 24 hours to aid in the comparisons. (Adapted from Calabrese and Kenyon 1991.)

Sometimes this stepwise strategy of choosing air pollution controls is very short-circuited and the designers and regulators go directly to the universal solution. One such approach is BACT (best available control technology). It is supposed that by utilizing the best amalgamation of filters, scrubbers and good work practices on the discharge source, a level of releases low enough to safeguard the most vulnerable target group would be accomplished (Hammer, 1999; An et al., 2011). Frequently, the subsequent level of emission will be quite below the minimum needed to safeguard the most vulnerable targets. This way all needless exposures must be eliminated. Examples of BACT are given in Table 5.4.

Table 5.4: Selected Examples of BACT (Best Available Control Technology) Displaying the Control Method Utilized and Estimated Efficiency

Process	**Pollutant**	**Control method**	**Estimated efficiency**
Soil remediation	Hydrocarbons	Thermal oxidizer	99
Production of fumed silica	Carbon monoxide	Good practice	50
Kraft pulp mill recovery boiler	Particulates	Electrostatic precipitator	99.68
Electric arc furnace	Particulates	Baghouse	100
Automobile painting	Hydrocarbons	Oven afterburner	90
Medical incinerator	Hydrogen chloride	Wet scrubber + dry scrubber	97.5
Petroleum refinery using catalytic cracking	Respirable particulates	Cyclone + Venturi scrubber	93
Waste disposal by dehydration and incineration	Particulates	Cyclone + condenser + Venturi scrubber + wet scrubber	95
Coal-fired boiler	Sulfur dioxide	Spray dryer + absorber	90
Asphalt plant	Hydrocarbons	Thermal oxidizer	99

BACT by itself doesn't guarantee adequate control levels. Even though BACT is the best control system centered on gas cleaning controls and good operating practices, BACT might not be very good if the source is the large plant or if it is situated next to the sensitive target. BACT (Best available control technology) must be tested to confirm that it is certainly good enough. The subsequent emission standards must be checked to conclude whether or not they might still be dangerous even with the finest gas cleaning controls. If the emission standards are still dangerous, other basic controls, like choosing safer materials or processes, or relocating in the less sensitive area, might have to be well-thought-out (Gundersen et al., 2006; Popp, 2006).

Another universal solution that evades some steps is the source performance standards. Several jurisdictions develop emission standards that can't be exceeded. Emission standards are centered on releases at the source. Normally this functions well, but like Best available control technology they can be untrustworthy. The levels must be low enough to preserve the maximum releases low enough to safeguard vulnerable target populations from the typical emissions. However, as with BACT, this might not be very good to safeguard everyone where there are huge discharge sources or nearby vulnerable populations. If this is the circumstance, other procedures should be used to guarantee the protection of all target groups (McHugh et al., 1997).

Both BACT and emission standards have a fundamental fault. They suppose that if certain standards are met at the plant, then the target groups will be protected automatically. This isn't essentially so, but once such a system is passed into the law, impacts on targets become secondary to defiance with the law (Mage et al., 1996; Guttikunda et al., 2014).

Source emission standards and BACT or the design criteria must be utilized as the minimum criteria for controls. If emission criteria or BACT will protect the vulnerable targets, then they can also be used as planned, else other administrative controls should be used (Yang et al., 2008).

5.3. CONTROL MEASURES

Controls can usually be divided into 2 basic kinds of controls – administrative and technological. Technological controls are described as the hardware which is put on the emission source to decrease pollutants in the gas stream to the level that is tolerable to the community and that will safeguard the most sensitive target. Administrative controls are described as other control measures (Ghose & Majee, 2001; Li et al., 2011; Lim et al., 2014).

5.3.1. Technological controls

Gas cleaning systems are usually placed at source, before the stack in order to eliminate pollutants from the gas stream before discharging it to the environment. Table 5.5 displays a brief summary of different classes of the gas cleaning system.

Table 5.5: Gas Cleaning Approaches for Eliminating Harmful Vapors, Gases, and Particulates From the Industrial Process Emissions

Control method	**Examples**	**Description**	**Efficiency**
Gases/Vapours			
Condensation	Contact condensers and surface condensers	The vapor is cooled and then condensed to a liquid. This is ineffective and is utilized as a pre-conditioner to other methods	80+% when concentration >2,000 ppm
Adsorption	Carbon Silica gel Alumina Molecular sieve	The gas or vapor is collected on the solid.	90+% when concentration <1,000 ppm 95+% when concentration >1,000 ppm
Absorption	Wet scrubbers	The gas or vapor is collected in the liquid.	82–95% when concentration <100 ppm 95–99% when concentration >100 ppm
Particulates			
Incineration	Flares Catalytic incinerator	An organic gas or vapor is oxidized by generally heating it to high temperature and holding at that temperature for an adequate time period.	Not recommended when concentration <2,000 ppm 80+% when concentration >2,000 ppm
Wet scrubbers	Wetted filter Venturi Tray or sieve scrubber	Liquid droplets gather the particles by interception, impaction, and diffusion. The droplets and the particles are then isolated from the gas stream.	For 5 μm particles, 98.5% at 6.8 w.g.; 99.99+% at 50 w.g. For 1 μm particles, 45% at 6.8 w.g.; 99.95 at 50 w.g.

Inertial separators	Cyclones	Particle-laden gases are enforced to change direction. The inertia of the particle triggers them to isolate from the gas stream. This is ineffective and is utilized as a pre-conditioner to other methods.	70–90%
Filters	Baghouse	A porous fabric eliminates particulates from the gas stream. The leaky dust cake that shapes on the fabric then essentially does the filtration.	99.9% for 0.2 μm particles 99.5% for 10 μm particles
Electrostatic precipitators	Plate-wire Flat-plate Tubular Wet	Electrical forces are utilized to transfer the particles out of gas stream onto the collection plates	95–99.5% for 0.2 μm particles 99.25–99.9+% for 10 μm particles

The gas cleaner is the part of a composite system comprising of hoods, fans, ductwork, cleaners, and stacks. The design, maintenance and performance of every part impact the performance of all the other parts and the whole system (Krupnick & Portney, 1991).

It should be observed that system efficiency changes widely for every kind of cleaner, depending on the energy input, design and the features of the gas stream and the pollutant. As an outcome, the sampling effectiveness in Table 5.5 is just approximations. The change inefficiencies are established with wet scrubbers in Table 5.5. Wet scrubber collection effectiveness goes from 98.5 % for 5 μm particles to 45 % for 1 μm particles at a similar pressure drop across the scrubber. For the similar size particle, 1 μm, effectiveness goes from 45% efficiency at 6.8–99.95 w.g at 0 w.g. As an outcome, gas cleaners should be matched to a particular gas stream in question. The utilization of generic devices isn't recommended (Fang & Chen, 1996; Zhang et al., 2016).

5.3.1.1. Waste disposal

When choosing and designing the gas cleaning systems careful consideration should be given to safe disposal of the gathered material. As displayed in Table 5.6, some procedures yield large amounts of pollutants. If most of the pollutants are gathered by the gas cleaning apparatus there can be a dangerous waste disposal problem (Seinfeld & Kyan, 1971; Okuda et al., 2011).

Table 5.6: Sample of Uncontrolled Emission Rates for Chosen Industrial Processes

Industrial source	**Emission rate**
100-ton electric furnace	257 tons/year particulates
41.7 ton/hr incinerator	208 lb/hr NO x
1,500 MM BTU/hr oil/gas turbine	444 lb/hr SO_2
100 trucks/day clear coat	3,795 lb/week organics

In some circumstances, the wastes might contain valued products that can be reprocessed, like heavy metals from the smelter or solvent from the painting line. The wastes can be utilized as the raw material for some other industrial process; for instance, sulfur dioxide gathered as sulfuric acid (H_2SO_4) can be utilized in the production of fertilizers.

Where the wastes can't be reused or recycled, disposal might not be simple. Not only the volume can be problematic, but they might be hazardous themselves. For instance, if the sulfuric acid (H_2SO_4) captured from a smelter or boiler can't be reused, it must be further treated in order to neutralize it before the disposal.

5.3.1.2. Dispersion

Dispersion can decrease the concentration of a contaminant at a target. However, it should be kept in mind that dispersion doesn't decrease the total volume of material leaving the plant. A tall stack only permits the plume to spread out and get diluted before reaching the ground level, where vulnerable targets are probable to exist. If the contaminant is mainly a nuisance, like an odor, dispersion might be acceptable.

Conversely, if the material is cumulative or persistent, like heavy metals, dilution might not be the solution to the air pollution problem (Schöpp et al., 1998).

5.3.2. Administrative Controls

Besides the technological systems, there exists another set of controls that should be well-thought-out in the complete design of an air pollution control system.

5.3.2.1. Substitution

One of the favorite occupational hygiene approaches for controlling environmental dangers in the place of work is to replace a safer process or material. If a safer material or process can be utilized, and dangerous emissions evaded, the efficacy or type of the controls becomes academic. It is best to avert the issue than it is to correct the bad first decision. Examples of substitution comprise the utilization of cleaner fuels, reduced temperatures in the dryers and covers for the bulk storage.

This applies to slight purchases along with the main design criteria for the plant. If only environmentally safe processes or products are purchased, there won't be any risk to the environment, outdoors or indoors. If inappropriate buying is done, the remainder of the program comprises struggling to compensate for that decision. If the low-cost but dangerous process or product is purchased it might require special handling equipment and procedures, and special disposal techniques. As an outcome, the cheap item might have only the low purchase price but the high price to utilize and dispose of it.

5.3.2.2. Local Ventilation

Controls are needed for all the recognized problems that can't be evaded by replacing safer methods or materials. Emissions begin at the individual workplace, not the stack. The ventilation system that controls and captures emissions at the source will aid to safeguard the community or individuals if it is correctly designed. The ducts and hoods of the ventilation system are normally part of the total air pollution control system.

Generally, a local ventilation system is desired. It doesn't dilute the pollutants and delivers the concentrated gas stream that is simpler to clean before discharge to the environment. Gas cleaning equipment is usually more effective when cleaning air with greater concentrations of pollutants. For instance, a capture hood over pouring spout of the metal furnace will avert pollutants from getting into the environment and transport the fumes to the gas cleaning system. In Table 5.5 it can be observed that cleaning efficiencies for adsorption and absorption cleaners increase with the concentration of the pollutant, and condensation cleaners aren't suggested for low levels of contaminants.

If pollutants aren't captured at the source and are permitted to escape through ventilation openings and windows, they become uncontrolled escapee emissions. In some situations, these uncontrolled emissions can have a substantial influence on the immediate neighborhood.

5.3.2.3. Isolation

Locating the plants away from vulnerable targets can be a chief control technique when engineering controls are insufficient by themselves. This might be the only way of accomplishing a tolerable level of control when BACT (best available control technology) should be relied on. If, after smearing the best accessible controls, a group of a target is still at risk, attention should be given to discovering a substitute site where sensitive populations aren't present.

5.3.2.4. Work Procedures

Work procedures should be developed to confirm that equipment is utilized safely and properly, without any risk to the workers or environment. Complex air pollution systems should be properly operated and maintained if they are desired to do their task as intended. A vital factor in this is staff training. Staff should be trained in how to utilize and maintain the equipment in order to reduce or remove the number of dangerous materials discharged to the community or workplace. In some situations, BACT is dependent on good practice in order to guarantee acceptable results.

5.3.2.5. Real-Time Monitoring

The system centered on real-time monitoring isn't popular and isn't commonly used. In this situation, meteorological monitoring and continuous emission can be joined with dispersion modeling to forecast downwind exposures. When the anticipated exposures approach the tolerable levels, the information is utilized to decrease production rates and emissions. This is an ineffective method, but might be a tolerable interim control technique for an existing facility.

The opposite of this is to broadcast warnings to the public when the conditions are such that extreme concentrations of pollutants might exist so that the public is able to take suitable action. For instance, if the warning is broadcasted that the atmospheric circumstances are such that the levels of sulfur dioxide downwind of the smelter are excessive, vulnerable populations like asthmatics would know that isn't favorable to go outside. Again, this might be a tolerable temporary control till permanent controls are installed.

Real-time meteorological and atmospheric monitoring is sometimes utilized to avert or decrease main air pollution events where several sources may exist. When it becomes obvious that the excessive levels of air pollution are likely, the personal utilization of cars might be restricted and the main emitting industries shut down.

5.3.2.6. Maintenance/Housekeeping

In all circumstances the efficiency of the controls is dependent on appropriate maintenance; the equipment must operate as intended. Not only should the controls of air pollution be maintained and utilized as intended, but the procedures generating potential emissions should be operated and maintained properly. An instance of an industrial procedure is the wood chip dryer with a failing temperature regulator; if the dryer is operated at high temperature, it will normally emit more materials and perhaps a distinct kind of material, from the drying wood. An instance of the gas cleaner maintenance disturbing releases would be a poorly maintained bag-house with wrecked bags, which would permit particulates to move through the filter (Heller et al., 1990; Mccrillis & Jaasma, 1993).

Housekeeping also plays a vital role in controlling the total emissions. Dusts that aren't rapidly cleaned up inside the plants can become re-entrained and are hazardous for the staff. If these dusts are carried properly outside of the plant, then they are a hazard for the community. Poor housekeeping in the plant yard could offer a substantial risk to the community. Uncovered bulk materials, vehicle-raised dusts or plant wastes can outcome in contaminants being carried on winds into the community. Keeping the yards clean, utilizing proper storage sites or containers, is significant in decreasing total emissions. A system should be not only properly designed but utilized properly as well if the whole community is to be secured (Graham & Lewellen, 1993; England et al., 2000).

5.4. EQUIPMENT FOR THE EMISSION SAMPLING

Source sampling can normally be carried out for many reasons (Jones & Nedwell, 1990; De Fre & Wevers, 1998):

1. To illustrate the emissions. To design the air pollution control system, one should know well what is being released or discharged. Not only the gas volume, but the identity, amount and, in the situation of particulates, should size distribution of material being released be known. The same information is essential to catalog total releases in a neighborhood.
2. To test equipment effectiveness. After the purchase of the air pollution control system, it must be tested to confirm that it is performing the intended job.
3. As part of the control system. When emissions are constantly observed, the data can be utilized to tune the air pollution control system or the plant process itself.
4. To decide compliance. When regulatory standards comprise the emission limits, emission sampling can be utilized to determine non-compliance or compliance with the standards.

The kind of sampling system utilized will be dependent on the purpose of taking the costs, samples, availability of technology and training of the staff (Wuske et al., 1998; Lee et al., 2004).

5.4.1. Visible Emissions

Where there is a need to decrease the soiling power of the air, enhance visibility or avert introduction of the aerosols into the atmosphere, standards might be centered on visible emissions (Richards & Johnson, 1976; Du & Yu, 2017).

Visible emissions are made up of small particles or colored gases. The denser a plume is the more material is emitted. This characteristic is apparent to the sight and trained spectators can be utilized to evaluate emission levels (Quickenden et al., 1982; Lin et al., 2003). There are many advantages to using this approach of evaluating emission standards (Mateos et al., 2005; yao et al., 2015):

1. No costly equipment is needed.
2. One person can make several observations in one day.
3. Plant operators can rapidly evaluate the effects of procedure variations at low cost.

4. Violators can be mentioned without timewasting source testing.
5. Questionable releases can be traced and the actual releases then determined by the source testing as defined in the following sections.

5.4.2. Extractive Sampling

A much more laborious sampling approach calls for the sample of the gas stream to be eliminated from the stack and examined. Although this looks simple, it doesn't translate into the simple sampling method.

The sample must be gathered iso-kinetically, specifically when particulates are collected (Elci et al., 2011). Isokinetic sampling is described as sampling by drawing the sample into the probe at a similar velocity that material is moving in a stack or the duct. This is normally done by measuring the gas stream velocity stream with the pitot tube and then fine-tuning the sampling rate to ensure that the sample moves in the probe at a similar velocity. This is necessary when sampling for the particulates, as heavier, larger particles won't follow the change in velocity or direction (Spicer et al., 2009; Cheng & Corporan, 2010). As an outcome concentration of the larger particles in the sample won't be illustrative of the gas stream and the sample will be incorrect.

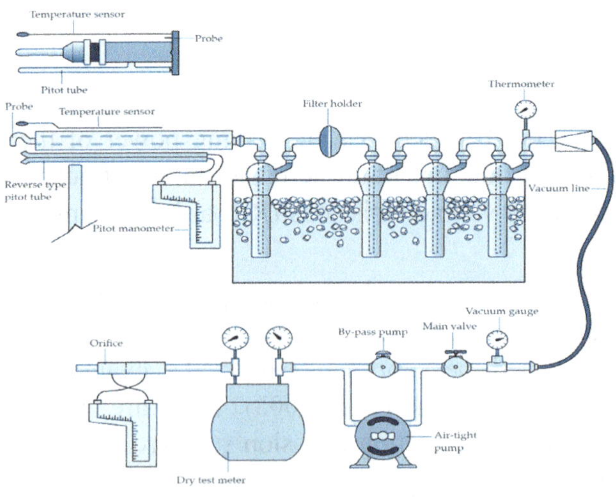

Figure 5.1: An illustration of isokinetic sampling train for SO_2 (sulfur dioxide).

[Source: http://www.ilocis.org/documents/chpt55e.htm]

The sampling train for SO_2 is displayed in Figure 5.1. It isn't simple and the trained operator needs to confirm that the sample is properly collected (Reyes et al., 2006; Johnson et al., 2016). If something other than SO_2 is to be sampled, impingers and the ice bath can be eliminated and the suitable collection device inserted. Extractive sampling, mainly isokinetic sampling, can be quite versatile and accurate, and has many uses (McDowell & Coleman, 1973; Escobar et al., 1998):

1. It is an acknowledged sampling method with sufficient quality controls, and therefore can be utilized to conclude compliance with standards.
2. The potential correctness of the technique makes it appropriate for performance testing of the new control equipment.
3. As samples can be gathered and examined under controlled laboratory circumstances for several components, it is helpful for describing the gas stream.

An automated and simplified sampling system can normally be connected to the continuous gas or particulate analyzer to constantly monitor emissions. This can deliver documentation of emissions and the rapid operating status of an air pollution control system (Agrawal & Sudhakar, 2002).

5.4.3. In Situ Sampling

Emissions can be sampled in the stack. Figure 5.2 is an illustration of the simple transmissometer utilized to measure the materials in the gas stream. In this instance, a light beam is anticipated across the stack to the photocell. The colored gas or particulates will block or absorb some of the light. The more the material is the less light will reach the photocell (Liu & Horlick, 1995; Aurell & Gullett, 2010).

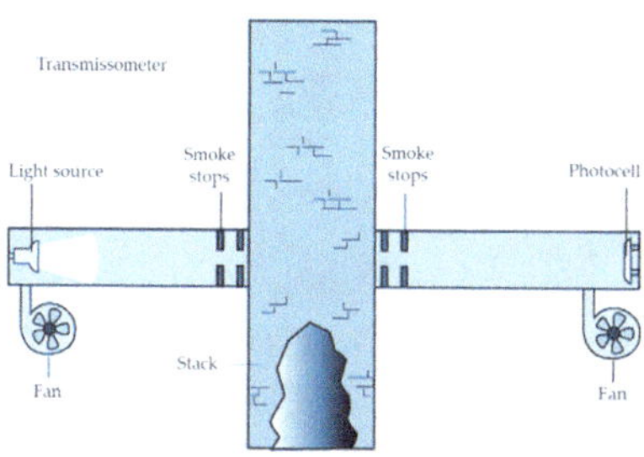

Figure 5.2: A simple transmissometer used measure particulates in the stack.

[Source: http://www.ilocis.org/documents/chpt55e.htm]

By using different detectors and light sources like ultraviolet light (UV), gases translucent to the visible light can be perceived. These devices can be fine-tuned to particular gases, and therefore can measure the concentration of gas in the waste stream (Matusiewicz et al., 1990; Lu et al., 2002).

This monitoring system has a benefit over the extractive system in that this can estimate the concentration across the complete stack or duct, while the extractive approach measures concentration only at a point from where the sample was removed. This can outcome in substantial error if the sample of the gas stream isn't well mixed. However, the extractive method provides more techniques of analysis, and therefore perhaps can be utilized in more applications (Sallé et al., 2005). As the in situ system offers a constant readout, it can be utilized to document releases or to adjust the operating system.

REFERENCES

1. Agrawal, Y. K., & Sudhakar, S. (2002). Extractive spectrophotometric and inductively coupled plasma atomic emission spectrophotometric determination of titanium by using dibenzo-18-crown-6. *Talanta, 57*(1), 97–104.
2. An, H. T. Q., Huu, T. P., Le Van, T., Cormier, J. M., & Khacef, A. (2011). Application of atmospheric non thermal plasma-catalysis hybrid system for air pollution control: Toluene removal. *Catalysis Today*, *176*(1), 474–477.
3. Aurell, J., & Gullett, B. K. (2010). Aerostat sampling of PCDD/PCDF emissions from the Gulf oil spill in situ burns. *Environmental Science & Technology*, *44*(24), 9431–9437.
4. Barbera, A. J., & McConnell, V. D. (1986). Effects of pollution control on industry productivity: a factor demand approach. *The Journal of Industrial Economics*, *1*, 161–172.
5. Cheng, M. D., & Corporan, E. (2010). A study of extractive and remote-sensing sampling and measurement of emissions from military aircraft engines. *Atmospheric Environment*, *44*(38), 4867–4878.
6. De Fre, R., & Wevers, M. (1998). Underestimation in dioxin emission inventories. *Organohalogen Compounds*, *36*(1199R), 17.
7. Du, P., & Yu, J. S. (2017). Near-ultraviolet light induced visible emissions in Er3+-activated La2MoO6 nanoparticles for solid-state lighting and non-contact thermometry. *Chemical Engineering Journal*, *327*, 109–119.
8. Elci, L., Kolbe, N., Elci, S. G., & Anderson, J. T. (2011). Solid phase extractive preconcentration coupled to gas chromatography–atomic emission detection for the determination of chlorophenols in water samples. *Talanta*, *85*(1), 551–555.
9. England, G. C., Zielinska, B., Loos, K., Crane, I., & Ritter, K. (2000). Characterizing PM2. 5 emission profiles for stationary sources: comparison of traditional and dilution sampling techniques. *Fuel Processing Technology*, *65*, 177–188.
10. Escobar, V. G., Tomé, F. V., Lozano, J. C., & Sánchez, A. M. (1998). Extractive procedure for uranium determination in water samples by liquid scintillation counting. *Applied Radiation and Isotopes*, *49*(7), 875–883.
11. Fang, S. H., & Chen, H. W. (1996). Air quality and pollution control in

Taiwan. *Atmospheric Environment, 30*(5), 735–741.

12. Ghose, M. K., & Majee, S. R. (2001). Air pollution caused by opencast mining and its abatement measures in India. *Journal of Environmental Management, 63*(2), 193–202.
13. Graham, M. M., & Lewellen, B. L. (1993). High-speed automated discrete blood sampling for positron emission tomography. *Journal of Nuclear Medicine: Official Publication, Society of Nuclear Medicine, 34*(8), 1357–1360.
14. Gundersen, P., Schmidt, I. K., & Raulund-Rasmussen, K. (2006). Leaching of nitrate from temperate forests effects of air pollution and forest management. *Environmental Reviews, 14*(1), 1–57.
15. Guttikunda, S. K., Goel, R., & Pant, P. (2014). Nature of air pollution, emission sources, and management in the Indian cities. *Atmospheric Environment, 95*, 501–510.
16. Hakanson, L. (1980). An ecological risk index for aquatic pollution control. A sedimentological approach. *Water Research, 14*(8), 975–1001.
17. Hammer, T. (1999). Application of plasma technology in environmental techniques. *Contributions to Plasma Physics, 39*(5), 441–462.
18. Heller, B., Klingenberg, H., Lach, G., & Winckler, J. (1990). Performance of a new system for emission sampling and measurement (SESAM). *SAE Transactions*, 126–136.
19. Johnson, M. R., Crosland, B. M., McEwen, J. D., Hager, D. B., Armitage, J. R., Karimi-Golpayegani, M., & Picard, D. J. (2016). Estimating fugitive methane emissions from oil sands mining using extractive core samples. *Atmospheric Environment, 144*, 111–123.
20. Jones, H. A., & Nedwell, D. B. (1990). Soil atmosphere concentration profiles and methane emission rates in the restoration covers above landfill sites: equipment and preliminary results. *Waste Management & Research, 8*(1), 21–31.
21. Junhua, L., Jianjun, C., & Jiming, H. (2005). Recent progress of air pollution control technology. *Chemical Industry and Engineering Progress, 24*(7), 703.
22. Krupnick, A. J., & Portney, P. R. (1991). Controlling urban air pollution: a benefit-cost assessment. *Science, 252*(5005), 522–528.
23. Lee, S. W., Herage, T., & Young, B. (2004). Emission reduction potential from the combustion of soy methyl ester fuel blended with

petroleum distillate fuel. *Fuel, 83*(11–12), 1607–1613.

24. Li, Y., Wang, W., Wang, J., Zhang, X., Lin, W., & Yang, Y. (2011). Impact of air pollution control measures and weather conditions on asthma during the 2008 Summer Olympic Games in Beijing. *International Journal of Biometeorology, 55*(4), 547–554.
25. Lin, H., Meredith, G., Jiang, S., Peng, X., Luo, T., Peyghambarian, N., & Pun, E. Y. B. (2003). Optical transitions and visible upconversion in Er 3+ doped niobic tellurite glass. *Journal of Applied Physics, 93*(1), 186–191.
26. Lin, H., Zhang, Y., Liu, T., Xiao, J., Xu, Y., Xu, X., ... & Ma, W. (2014). Mortality reduction following the air pollution control measures during the 2010 Asian Games. *Atmospheric Environment, 91*, 24–31.
27. Liu, X. R., & Horlick, G. (1995). In situ laser ablation sampling for inductively coupled plasma atomic emission spectrometry. *Spectrochimica Acta Part B: Atomic Spectroscopy, 50*(4–7), 537–548.
28. Long, N. V. (1992). Pollution control: A differential game approach. *Annals of Operations Research, 37*(1), 283–296.
29. Lu, X., Faguy, P. W., & Liu, M. (2002). In Situ Potential-Dependent FTIR Emission Spectroscopy A Novel Probe for High Temperature Fuel Cell Interfaces. *Journal of the Electrochemical Society, 149*(10), A1293-A1298.
30. Lübbe-Wolff, G. (2001). Efficient environmental legislation-On different philosophies of pollution control in Europe. *Journal of Environmental Law, 13*(1), 79–87.
31. Mage, D., Ozolins, G., Peterson, P., Webster, A., Orthofer, R., Vandeweerd, V., & Gwynne, M. (1996). Urban air pollution in megacities of the world. *Atmospheric Environment, 30*(5), 681–686.
32. Mateos, X., Solé, R., Gavaldà, J., Aguiló, M., Díaz, F., & Massons, J. (2005). Ultraviolet and visible emissions of Er3+ in KY (WO4) 2 single crystals co-doped with Yb3+ ions. *Journal of luminescence, 115*(3–4), 131–137.
33. Matusiewicz, H., Sturgeon, R. E., & Berman, S. S. (1990). In situ hydride generation preconcentration of arsenic in a graphite furnace with sample vaporization into a microwave induced plasma for emission spectrometry. *Spectrochimica Acta Part B: Atomic Spectroscopy, 45*(1–2), 209–214.

34. Mccrillis, R. C., & Jaasma, D. R. (1993). Woodstove emission measurement methods: comparison and emission factors update. *Environmental Monitoring and Assessment*, *24*(1), 1–12.
35. McDowell, W. J., & Coleman, C. F. (1973). Some methods of controlling quenching in extractive scintillators for liquid scintillation counting of alpha-emitting nuclides. *Analytical Letters*, *6*(9), 795–799.
36. McHugh, C. A., Carruthers, D. J., & Edmunds, H. A. (1997). ADMS–Urban: an air quality management system for traffic, domestic and industrial pollution. *International Journal of Environment and Pollution*, *8*(3–6), 666–674.
37. Milford, J. B., Russell, A. G., & McRae, G. J. (1989). A new approach to photochemical pollution control: Implications of spatial patterns in pollutant responses to reductions in nitrogen oxides and reactive organic gas emissions. *Environmental Science & Technology*, *23*(10), 1290–1301.
38. Okuda, T., Matsuura, S., Yamaguchi, D., Umemura, T., Hanada, E., Orihara, H., ... & Liang, L. (2011). The impact of the pollution control measures for the 2008 Beijing Olympic Games on the chemical composition of aerosols. *Atmospheric Environment*, *45*(16), 2789–2794.
39. Pattrick, G., Van der Lingen, E., Corti, C. W., Holliday, R. J., & Thompson, D. T. (2004). The potential for use of gold in automotive pollution control technologies: a short review. *Topics in Catalysis*, *30*(1–4), 273–279.
40. Penetrante, B. M. (1993). Pollution control applications of pulsed power technology. In *Ninth IEEE International Pulsed Power Conference* (Vol. 1, June, p. 1–22). IEEE.
41. Popp, D. (2006). International innovation and diffusion of air pollution control technologies: the effects of NOX and SO_2 regulation in the US, Japan, and Germany. *Journal of Environmental Economics and Management*, *51*(1), 46–71.
42. Quickenden, T. I., Trotman, S. M., & Sangster, D. F. (1982). Pulse radiolytic studies of the ultraviolet and visible emissions from purified H2O ice. *The Journal of Chemical Physics*, *77*(8), 3790–3802.
43. Reyes, F., Grutter, M., Jazcilevich, A., & González-Oropeza, R. (2006). Tecnical Note: Analysis of non-regulated vehicular emissions by extractive FTIR spectrometry: tests on a hybrid car in Mexico City.

Atmospheric Chemistry and Physics, *6*(12), 5339–5346.

44. Richards, J. L., & Johnson, P. M. (1976). The visible emissions of molecular oxygen in rare gas solids. *The Journal of Chemical Physics*, *65*(10), 3948–3951.
45. Sallé, B., Cremers, D. A., Maurice, S., Wiens, R. C., & Fichet, P. (2005). Evaluation of a compact spectrograph for in-situ and stand-off laser-induced breakdown spectroscopy analyses of geological samples on Mars missions. *Spectrochimica Acta Part B: Atomic Spectroscopy*, *60*(6), 805–815.
46. Schöpp, W., Amann, M., Cofala, J., Heyes, C., & Klimont, Z. (1998). Integrated assessment of European air pollution emission control strategies. *Environmental Modelling & Software*, *14*(1), 1–9.
47. Seinfeld, J. H., & Kyan, C. P. (1971). Determination of optimal air pollution control strategies. *Socio-Economic Planning Sciences*, *5*(3), 173–190.
48. Spicer, C. W., Holdren, M. W., Cowen, K. A., Joseph, D. W., Satola, J., Goodwin, B., … & Newburn, M. (2009). Rapid measurement of emissions from military aircraft turbine engines by downstream extractive sampling of aircraft on the ground: Results for C-130 and F-15 aircraft. *Atmospheric Environment*, *43*(16), 2612–2622.
49. Tietenberg, T. (1995). Tradeable permits for pollution control when emission location matters: what have we learned?. *Environmental and Resource Economics*, *5*(2), 95–113.
50. Tietenberg, T. (2002). 19 Lessons from using transferable permits to control air pollution in the United States. *Handbook of Environmental and Resource Economics*, *275*, 1–20.
51. Wong, T. H., & Somes, N. L. G. (1995). A stochastic approach to designing wetlands for stormwater pollution control. *Water Science and Technology*, *32*(1), 145.
52. Wuske, T., Fittkau, I., Mahn, J., Polzius, R., & Manns, A. (1998). Pentachlorophenol detection at the source of emission: Sampling equipment and immunochemical analysis. *Analytica Chimica Acta*, *359*(3), 321–328.
53. Yang, J., Yu, Q., & Gong, P. (2008). Quantifying air pollution removal by green roofs in Chicago. *Atmospheric Environment*, *42*(31), 7266–7273.

54. Yao, D., Liu, H., Liu, Y., Dong, C., Zhang, K., Sheng, Y., … & Yang, B. (2015). Phosphine-free synthesis of Ag–In–Se alloy nanocrystals with visible emissions. *Nanoscale*, *7*(44), 18570–18578.
55. Zhang, H., Wang, S., Hao, J., Wang, X., Wang, S., Chai, F., & Li, M. (2016). Air pollution and control action in Beijing. *Journal of Cleaner Production*, *112*, 1519–1527.

CHAPTER 6

Water Pollution Control

CONTENTS

6.1. INTRODUCTION

The significance of water for sustenance of life cannot be overemphasized. Regardless of whether it is being used of running water in our homes, raising steers and developing harvests in our farms, or the expanded uses in industry, stay limitless. It is significant in this way, to not excessive consumption of this item either through tainting, or indiscreet use brings about genuine outcomes (Musa, 2013).

Water is viewed as contaminated if a few substances or condition is available to such an extent, that the water cannot be utilized for a particular reason. Olaniran (1995) characterized water contamination to be the existence of over the top measures of a danger (poisons) in water so that it is no longer appropriate for drinking, washing, cooking or different uses. Contamination is the introduction of pollution into nature (Webster.com, 2010). It is made by commercial and industrial waste, farming practices, regular human exercises and most eminently, models of transportation. Regardless of where you go and what you do, there are remainders earth's natural and it's occupants from numerous points of view. The three primary kinds of contamination are Land Contamination, Air Contamination and Water Contamination. Both with the end goal of this exploration, accentuation is on water contamination and control (Nzewi et al., 1985).

6.2. SOURCES OF WATER POLLUTION

Water pollution in Nigeria arises from various activities, among which are:

1. Sewage leakages
2. High population density
3. Oil spillage
4. Building visionaries and lavatories over running water or even the sea as it the repetition in some riverine parts
5. Overflowing during the rainy season which transmits waste deposits into our waters
6. Contamination of groundwater through boring events
7. Heavy metal
8. Water hyacinth and Menace of Nipa palm
9. Industrial waste dumped into our waters

10. Radioisotopes
11. Combustion
12. Eroded sediments
13. Littering
14. Herbicides and fertilizers
15. Household chemicals
16. Failing septic system
17. Toxic waste disposal at sea
18. Mineral processing plant (e.g., coal production)
19. Deforestation
20. Mining
21. Pesticides
22. Animal wastes
23. Herbicides and fertilizers

Water contamination is commonly actuated by people. It results from the activities of people carried on to a better self. These could be treated under the different exercises that man takes part in, which lead to contamination. The development of the human populace, agricultural and industrial practices is a significant reason for contamination. Water contamination turns out to be more awful because of congestion in urban territories. Horticultural, local and modern squanders are the significant contaminations of agnatic natural surroundings. Sewage is the greatest contamination of freshwater when released into them. Sewage is the waterborne waster of society and the release of untreated sewage into a stream is huge and undesirable. The striking result is a substantial and quick drop in the measure of dissolved oxygen in the water. This happens on the grounds that organic matter invigorates decomposers particularly microscopic organisms that separate suspended solids in the sewage. As they breathe, the decomposers go through dissolved oxygen (O_2) and the Organic Oxygen Demand (Body) decreases. The verdure of the streams experiences change and decrease in number because of death by suffocation.

Exceptionally contaminated waterways have an offensive smell and contain practically zero vegetation or fauna. Another wellspring of water contamination is the release of high temp water from cooling motors in the industries. This expands water temperature and brings down the metabolic pace of living beings. This at that point raises their oxygen demand. The impacts of contamination are more noteworthy in shallow, encased or moderate streaming streams. Abundance compost, herbicides and pesticides when washed by downpour into waterways makes serious threat life. Abundance phosphorus in manure causes genuine eutrophication. Aside from fertilizers, detergent is likewise extremely dangerous to marine life when washed into the water. Chemical poisons from distaffs have been seen as animal cancer-causing agents. The coloring enterprises in Nigeria (tie and color) produce synthetic compounds, for example, copper salts and zinc sulfate which are non-biodegradable, when they are released into streams; they produce obliterating impacts on aquatic environments (Simeon and Ambah, 2013; Halder and Islam, 2015).

Contamination poses a critical hazard to life, particularly when the water is a wellspring of drinking and for household purposes for people dirtied waters are powerful operators of maladies, for example, cholera, tuberculosis and typhoid. A significant water toxin has been oil spilled in enormous amounts from tankers of broken oil pipes from oil industries that executes ocean weeds, marine winged animals, scavengers, mollusks, fishes and other ocean life forms that fill in as nourishment for people. This prompts calcium deficiencies in our eating routine. A few bug sprays like DDT are especially dangerous when permitted into waterways since its focus increments along with the natural pecking order. Clams, for example, can aggregate DDT to a focus, multiple times that of DDT in ocean water. The impacts of water contamination in certain zones has been to a degree of irreversibly changing amphibian biological systems. This is hazardous to plants and animals including people (Kaze and Gam, 2013).

Since water contamination has direct results on human health, a viable instructing methodology in the conventional training area is fundamental for a superior understanding in order to build up the correct frame of mind towards water (Aliyu et al., 2013). This is the reason the guided revelation approach is a showing system which when enough used and joined with different strategies for science training will leave enduring impact on the student just as assist him with taking care of the issues of his prompt condition.

6.3. EFFECTS OF WATER POLLUTION

Water contamination has a duel impact on nature. It effectively affects the living and furthermore on nature. The impacts of contamination on individuals and oceanic communities are numerous and varied. Water contamination causes around 14,000 deaths each day, generally because of the contamination of drinking water by untreated sewage in developing nations (Tim and Jaunty, 1994; Owa, 2013). An expected 700 million Indians have no access to an appropriate toilet, and 1,000 Indians youngsters die of diarrhea each day thus numerous different nations as well. About 500 million Chinese need access to safe drinking water (Novotny, 1994; Dosskey, 2001).

Certainly, with all these, we can anticipate that there should be a decrease in profitability. Biomass and a decent variety of communities are not out of the ordinary when an enormous amount of dangerous materials are discharged into the streams, lakes and seaside waters in the sea. A lot of amphibian contamination includes sewage in which natural waste prevails (Omernik and Griffith, 1991). This waste can expand auxiliary efficiency while modifying the character of the amphibian community. Most fishes particularly the species wanted as nourishment by man are among the delicate species that vanish with the least extreme contamination (Novotny, 1994; Chan and Huang, 2003).

Water contamination prompts harm to human wellbeing. Malady conveying agents, for example, microorganisms and infections are conveyed into the surface and groundwater. Drinking water is influenced and health risks. Direct harm to plants and animal nourishment additionally influences human wellbeing. Plants supplements including phosphorus, nitrogen and different substances that help the development of aquatic plant life could be in abundance causing algal agony and unreasonable weed development (Volk et al., 2008; Yang and Wang, 2010). This makes water to have scent, taste and in some cases shading. At last, the natural equalization of a waterway is adjusted. Nitrogen oxides and sulfur dioxide cause corrosive downpour which brings down the pH estimation of soil and discharge of carbon dioxide causes sea acidification, the continuous decline in the PH of the Earth's Seas as CO2 gets dissolved (Tamminga, 1992; Liding and Bojie, 2000).

6.4. POLLUTION MANAGEMENT AND CONTROL

There are numerous methodologies that could be embraced in water contamination control and treatment. It could be through anticipation, practice endeavors or join a venture/program; Guideline and checking or taking part in charge quantifies by lessening or limiting waste. Counteraction of water contamination as indicated by Wikipedia incorporates the following ways (Youthful et al., 1989; Puckett, 1995):

1. Wash your vehicle far away from any stormwater channels.
2. Don't throw junk, synthetic chemicals or solvents into sewer channels.
3. Inspects your septic framework each 3–5 years.
4. Avoid utilizing pesticides and fertilizers that can run off into water frameworks.
5. Sweep your garage as opposed to hosing it down.
6. Always siphon your waste-holding tanks on your vessel.
7. Use non-dangerous cleaning materials.
8. Clean up oil and other fluid spills with kitty litter and clear them up.
9. Do not wash paintbrushes in the sink.

Another method is to join or engage with contamination avoidance is to rehearse endeavors without anyone else or join ventures or program. A portion of these is accessible with the Environmental Protection Agency website (EPA).

Guideline and checking are a successful method for the contamination of the executives. Numerous countries worldwide have sanctioned enactment to manage different kinds of contamination just as to relieve the unfavorable impacts of contamination.

Contamination control intends to control the outflows and effluents into the air, water or soil and land. Without contamination control, the waster items from utilization, warming, agribusiness, mining, manufacturing, transportation, and other human activities, regardless of whether they gather or scatter, will debase the earth. Contamination anticipation and waste minimization are more needed than contamination control. In any case, contamination could be limit by embracing these practices (i) by reusing, (ii) by reusing, (iii) waste minimization, (iv) by moderating, (v) by forestalling, and (vi) by compost.

Aside from all these referenced above, you can likewise utilize contamination control gadgets which incorporate Dust collection system for example pack houses, electrostatic precipitators, cyclones, scrubbers, for example, ejector venturi scrubber, baffle spray scrubber, wet scrubber, splash tower, precisely helped scrubbers, sewage treatment, for example, sedimentation (essential treatment), actuated sludge bio channels (optional treatment, additionally utilized for modern wastewater), circulated air through tidal ponds, built wetlands (likewise utilized in urban spillover);

industrial wastewater treatment for example ultra-filtration, Programming interface oil-water separators, bio channels, dissolved air flotation (DAF), powdered actuated carbon treatment; the last however not the least is fume recuperation system and phytoremediation (Characklis and Wiesner, 1997; Borah and Bera, 2003).

6.5. HYDROLOGIC WATERS

The world's common waters might be seen as a ceaselessly flowing system as appeared in Figure 6.1, which gives a realistic delineation of waters in the hydrologic cycle, containing both surface and subsurface waters (Hagedorn et al., 1999; Kaushal et al., 2011).

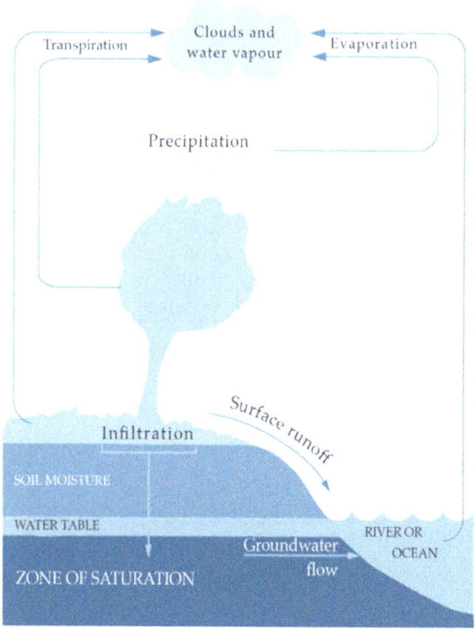

Figure 6.1: The hydrologic cycle.

[Source: http://www.ilocis.org/documents/chpt55e.htm]

As a kind of perspective for water quality, refined waters (H2O) presents the most noteworthy condition of purity. Waters in the hydrologic cycle might be seen as normal, however, they are not unadulterated. They become dirtied from both natural and human actions. Natural degradation impacts may result from a heap of sources – from fauna, verdure, lightning strikes causing fires, volcano eruptions, etc., which on a long haul premise are viewed as prevailing background levels for logical purposes (Bashiri et al., 2013a, b; Rajovic and Bulatovic, 2013).

Human-made contamination upsets the regular parity by superimposing waste materials released from different sources. Contamination might be brought into the waters of the hydrologic cycle anytime. For instance: barometrical (precipitation) may get debased via air toxins; surface waters may get dirtied in the spillover procedure from watersheds; sewage might be released into streams and waterways; and groundwater may get dirtied through invasion and underground contamination (Grema et al., 2013; Oladunjoye et al., 2013).

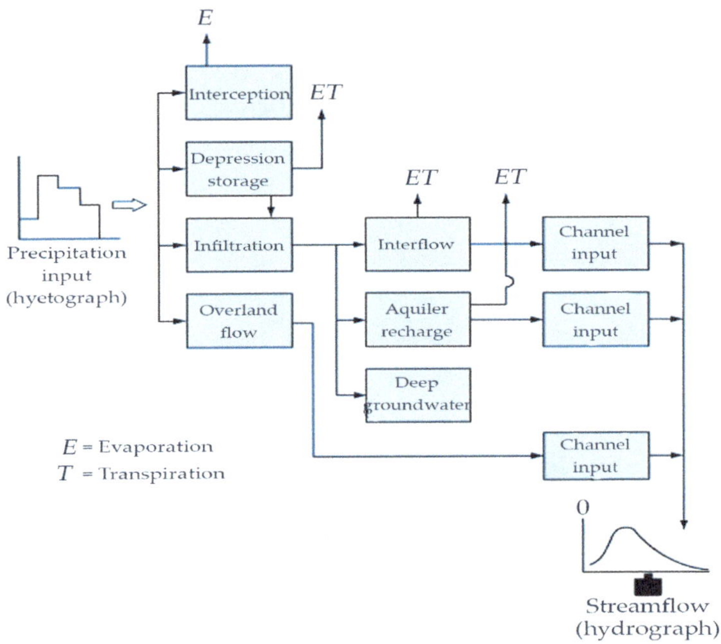

Figure 6.2: Distribution of precipitation.

[Source: http://www.ilocis.org/documents/chpt55e.htm]

Figure 6.2 shows an appropriation of hydrologic waters. Contamination is then superimposed on these waters and may in this way be seen as an unnatural or unequal ecological condition (Alamu, 2013; Nwona, 2013). The procedure of contamination may happen in waters of any piece of the hydrologic cycle and is progressively clear on the earth's surface as spillover from watersheds into streams and waterways. Anyway, groundwater contamination is additionally of a major ecological effect and is talked about after the segment on surface water contamination (Ellis, 1937; Chutter, 1972).

6.6. WATERSHED SOURCES OF WATER POLLUTION

Watersheds are the beginning space of surface water contamination. A watershed is characterized as a zone of the earth's surface on which hydrologic waters fall, gather, are utilized, discarded, and in the long run, are released into streams, waterways or different waterways (Chukwu et al., 2013; Obilonu et al., 2013). It is composed of a seepage system with extreme spillover or collection in a stream or waterway. Enormous stream watersheds are generally alluded to as seepage bowls. Figure 6.3 is a portrayal of the hydrologic cycle on a territorial watershed. For an area, the air of the different waters can be composed as a basic condition, which is the essential condition of hydrology; the run of the mill units are mm/year: P – R – G – E – T = ±S,

where: P = precipitation (i.e., rainfall, snowfall, hail); R = runoff or watershed surface flow; G = groundwater; E = evaporation.

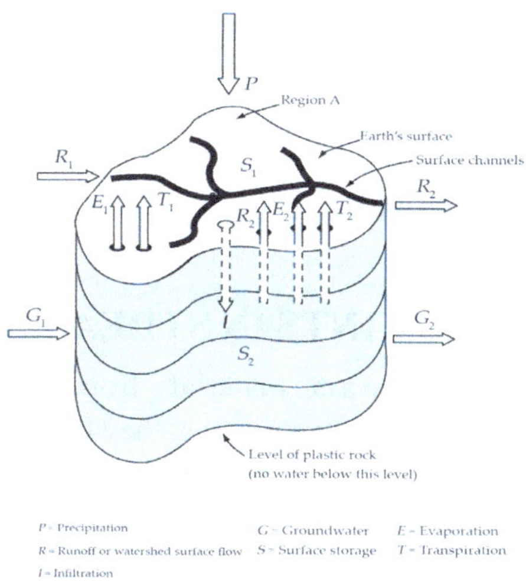

T = transpiration

S = surface storage

Figure 6.3: Regional hydrologic cycle.

[Source: http://www.ilocis.org/documents/chpt55e.htm]

Precipitation is seen as the starting structure in the above hydrologic plan. The term spillover is synonymous with the stream. Capacity alludes to supplies or detainment system which gather waters; for instance, a human-made dam (flood) on a stream makes a repository for motivations behind water stockpiling. Groundwater gathers as a storage system and may spill out of one area to another; it might be influent or gushing in connection to surface streams. Dissipation is a water surface marvel, and transpiration is related to transmission from biota (Loya, 1975; Loos et al., 2009).

In spite of the fact that watersheds may differ extraordinarily in size, certain seepage systems for water contamination assignment are named urban or non-urban (farming, undeveloped, rural) in character. Contamination happening inside these waste system starts from the accompanying sources:

Point sources: waste releases into a receiving water body at a particular area, at a point, for example, a sewer funnel or some sort of concentrated system outlet (Kahn and Rubin, 1989; Hines, 1966).

Non-point (scattered) sources: contamination entering a receiving water body from scattered sources in the watershed; uncollected precipitation spillover water waste into a stream is average. Non-point sources are additionally now and then alluded to as "diffuse" waters; notwithstanding, the term scattered is viewed as progressively enlightening.

Irregular sources: from a point or source which releases in specific situations, for example, with over-burden conditions; consolidated sewer floods during substantial precipitation overflow periods are regular (Rathbun and Tai, 1981; Sliva and Williams, 2001).

6.7. WATER POLLUTANTS IN STREAMS AND RIVERS

At the point when harmful waste materials from the above sources are released into streams or different waterways, they become poisons that have been characterized and portrayed in an earlier section (Lindegaard, 1995; Karr and Dudley, 1981). Toxins or contaminants which enter a waterway can be additionally partitioned into:

1. Degradable (non-preservationist) toxins: polluting influences which in the end break down into harmless substances or which might be evacuated by treatment strategies; that is, certain natural materials and synthetic compounds, household sewage, heat, plant supplements, most microscopic organisms and infections, certain dregs (Ringquist, 1994; Karr and Chu, 2000).

2. Non-degradable (preservationist) toxins: polluting influences that exist in the water condition and do not diminish in concentration except if weakened or expelled through treatment; that is, certain natural and inorganic synthetic substances, salts, colloidal suspensions (Chugh et al., 1978).
3. Hazardous waterborne poisons: complex types of harmful wastes including poisonous trace metals, certain inorganic, and natural Compounds.
4. Radionuclide poisons: materials that have been exposed to a radioactive source.

6.8. WATER POLLUTION CONTROL REGULATIONS

Extensively applicable water contamination control guidelines are commonly proclaimed by national administrative organizations, with increasing point by point guidelines by states, areas, regions, water regions, conservation locale, sanitation commissions and others. At the national and state (or region) levels, environmental protection agencies (EPAs) and services of health are typically charged with this duty. In the discourse of guidelines underneath, the configuration and certain parts follow the case of the water quality measures at present appropriate for the US State of Ohio.

6.8.1. Water Quality Use Designations

A definitive objective in the control of water contamination would be zero release of poisons to water bodies; notwithstanding, a complete accomplishment of this goal is generally not practical. The favored methodology is to set confinements on waste disposal releases for the sensible security of human well-being and the environment (Smith et al., 1983; Yoder and Rankin, 1998). In spite of the fact that these measures may fluctuate generally in various locales, use assignments for explicit waterways are normally the premise, as briefly discussed underneath.

6.8.1.1. Water Supplies

1. Public water supply: waters with regular treatment will be appropriate for human utilization.
2. Agricultural stockpile: waters appropriate for the water system and domesticated animals.

3. Industrial/business supply: waters reasonable for commercial and industrial uses with or without treatment.

6.8.1.2. Recreational Activities

1. Bathing waters: waters which during specific seasons are appropriate for swimming as affirmed for water quality alongside defensive conditions and offices.
2. Primary contact: waters which during specific seasons are appropriate for full-body contact entertainment, for example, swimming, kayaking and submerged plunging with negligible danger to public health because of water quality.
3. Secondary contact: waters which during specific seasons are appropriate for halfway body contact entertainment, for example, yet not constrained to, swimming, with insignificant danger to public health because of water quality.

Open water assets are classified as water bodies that exist in-park systems, wetland, natural life zones, wild, picturesque and recreational waterways and freely possessed lakes, and waters of excellent recreational or environmental importance (Yoder et al., 1999).

6.8.1.3. Aquatic Life Habitats

Normal designations will shift as indicated by atmospheres, yet identify with conditions in water bodies for supporting and keeping up certain amphibian life forms, particularly different types of fish. For instance, use designations in a calm atmosphere as subdivided in guidelines for the State of Ohio Environmental Protection Agency (EPA) are listed underneath without description:

1. Warm water
2. Limited warm water
3. Exceptional warm water
4. Modified warm water
5. Seasonal salmonid
6. Coldwater
7. Limited resource water.

6.8.2. Water Pollution Control Criteria

Wastewater and natural waters and are described by their biological, physical and chemical composition. The essential physical properties and the biological and chemical constituents of wastewater and their sources are a protracted rundown.

Each assigned water body should be controlled by guidelines that might constitute both essential and increasingly definite numerical criteria as briefly shown below (Wilhm and Dorris, 1968; Wu et al., 2005). Fundamental opportunity from contamination. To the degree useful and conceivable, all waterways ought to achieve the essential criteria of the "Five Opportunities from Contamination":

1. Free from suspended solids or different substances that enter the waters because of human action and that will settle to frame foul or otherwise shocking sludge stores, or that will unfavorably influence aquatic life.
2. Free from coasting garbage, oil, filth and other skimming materials entering the waters because of human movement in sums adequate to be unattractive or cause corruption.
3. Free from materials entering the waters because of human movement, creating shading, smell or different conditions in such degree as to create an irritation.
4. Free from substances entering the waters because of human movement, in focuses that are poisonous or destructive to human, aquatic life as well as are quickly deadly in the blending zone.
5. Free from supplements entering the waters because of human movement, in focuses that make annoyance developments of oceanic weed sand green growth.

Water quality criteria are numerical constraints and rules for the control of chemical, organic and dangerous constituents in waterways (Palange and Zavala, 1987; Wood, 1995).

With more than 70,000 or more chemical mixes being used today it is unrealistic to determine the control of each. In any case, criteria for synthetic compounds can be built up based on constraints as they as a matter of first importance identify with three significant classes of utilization and introduction:

Class 1: Chemical criteria for security of human well-being are of first significant concern and ought to be set by proposals from legislative well-being offices, the WHO and perceived well-being research associations.

Class 2: Synthetic criteria for control of rural water supply ought to be founded on perceived logical investigations and suggestions which will ensure against unfriendly consequences for yields and domesticated animals because of a harvest water system and domesticated animals watering.

Class 3: Chemical criteria for security of aquatic life should be founded on perceived logical examinations in regards to the affect-ability of these species to explicit synthetics and furthermore as identified with human utilization of fish and ocean nourishment (Stahl Jr, 1991).

Wastewater profluent criteria identify with impediments on poison constituents present in wastewater effluents and are a further strategy for control. They might be set as identified with the water use assignments of waterways and as they identify with the above classes for substance criteria (Byron and Goldman, 1989; Rankin, 1995). Biological criteria depend on water body environment conditions which are expected to help aquatic life.

6.8.3. The Organic Content of Wastewaters and Natural Waters

The gross substance of the organic matter is generally significant in portraying the pollution quality of both natural waters and wastewater (Stevens, 1966; Liquete et al., 2016).

Three research facility tests are commonly utilized for this reason:

1. Biochemical oxygen demand (BOD): A five-day Body (BOD5) is the most broadly utilized parameter; this test gauges the dissolved oxygen utilized by micro-organisms in the biochemical oxidation of natural matter over this period (Leenheer, 1981; Mama et al., 2001).
2. Chemical oxygen demand (COD): this test is to gauge the natural matter in industrial wastes and metropolitan that contain exacerbates that are harmful to natural life; it is a proportion of what could be compared to the organic matter that can be oxidized (Imai et al., 2002; Galambos et al., 2004).
3. Total organic carbon (TOC): this test is particularly suitable to little amounts of organic matter in water; it is a measure of the

organic matter that is oxidized to carbon dioxide (Najm and Trussell, 2001; Mitch et al., 2003).

6.8.4. Antidegradation Policy Regulations

Antidegradation policy rules are a further method for stopping the spread of water contamination outside certain usual circumstances (Kimes, 1995; Harleston, 1996). As an instance, the Ohio Environmental Protection Agency Water Quality Standards antidegradation arrangement comprises of three levels of assurance (Chilson, 1998; Brawer, 1999):

Level 1: Existing users must be kept up and ensured. No further water quality debasement is permitted that would meddle with existing assigned employments.

Level 2: Next, water quality superior to anything that expected to ensure utilizes must be kept up except if it is indicated that lower water quality is fundamental for significant monetary or social improvement, as controlled by the EPA Executive (Hersh, 2009).

Level 3: At last, the nature of water asset waters must be kept up and ensured. Their current encompassing water quality is not to be debased by any substances resolved to be harmful or to meddle with any assigned use. Expanded poison loads are permitted to be released into water bodies in the event that they do not bring about lowering down existing water quality (Swihart et al., 1986).

6.8.5. Water Pollution Discharge Mixing Zones and Waste Load Allocation Modeling

Blending zones are territories in a waterway that take into account treated or untreated wastewater releases to achieve balanced out conditions, as outlined in Figure 6.4 for a streaming stream. The release is at first in a transient state which turns out to be continuously weakened from the source concentration to the getting water conditions. It is not to be considered as a treatment element and might be depicted with explicit limitations (Dillon and Rigler, 1974; Griffith et al., 1994).

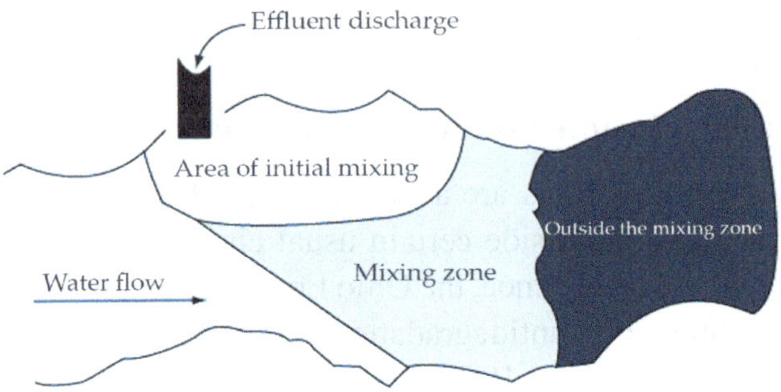

Figure 6.4: Mixing zones.

[Source: http://www.ilocis.org/documents/chpt55e.htm]

Typically, mixing zones must not:

1. Interfere with migration, survival, reproduction or growth of aquatic species
2. Include spawning or nursery areas
3. Include public water supply intakes
4. Include bathing areas
5. Establish more than 1/2 the width of a stream
6. Establish more than 1/2 the cross-sectional area of a stream mouth
7. Encompass downstream for a distance more than five times the stream width.

Waste load portion examines have become significant on account of the significant expense of supplement control of wastewater releases to maintain a strategic distance from in-stream eutrophication (characterized underneath). These investigations, for the most part, utilize the PC models for simulation of water quality conditions in a stream, especially concerning supplements, for example, types of phosphorus and nitrogen, which influence the dissolved oxygen elements (Gordon, 1985; Deflandre et al., 2006). Conventional water quality models of this sort are spoken to by the US EPA model QUAL2E. A later model is the Omni Diurnal Model (ODM), which incorporates a simulation of the effect of established vegetation on an in-stream supplement and dissolved oxygen dynamics.

6.8.6. Variance Provisions

All water contamination control guidelines are restricted in flawlessness and along these lines should incorporate arrangements which take into account critical difference dependent on specific conditions which may forestall quick or complete consistency (Crandall et al., 1999. Rajinikanth et al., 2009).

6.9. RISK ASSESSMENT AND MANAGEMENT AS RELATED TO WATER POLLUTION

The above water contamination control guidelines are an average of worldwide legislative methodologies for accomplishing consistency with water quality benchmarks and wastewater pro-fluent release limits. By and large, these guidelines have been determined to the premise of health factors and logical research; where some vulnerability exists as to potential impacts, security factors regularly are applied. The usage of some of these guidelines might be absurd and exceedingly expensive for the general population everywhere just as for private endeavors (Tchounwou et al., 2004; Lewis et al., 2016). In this manner, there is a developing worry for increasingly productive designation of resources in accomplishing objectives for water quality improvement. As recently brought up in the talk of hydrologic waters, immaculate purity does not exist even in naturally occurring waters (Huang and Xia, 2001; Grifoll et al., 2013).

A developing innovative methodology supports appraisal and the executives of biological dangers in the setting of water contamination guidelines. The idea depends on an investigation of the biological advantages and expenses in satisfying guidelines or breaking points. Specialists have proposed the use of oceanic environmental hazard evaluation as a guide in setting water contamination control limits, especially as appropriate for the assurance of aquatic life (Safe, 1998; James and Joyce, 2004). Such hazard appraisal techniques might be applied to assess the environmental impacts of chemical concentrations for a wide scope of surface water contamination conditions including:

1. Point source pollution.
2. Non-point source pollution.
3. Existing polluted sediments in stream channels.
4. Dangerous wastes sites as linked to water bodies.

5. Examination of present water pollution control standards.

The suggested method contains three tiers; as shown in Figure 6.5 which demonstrates the method.

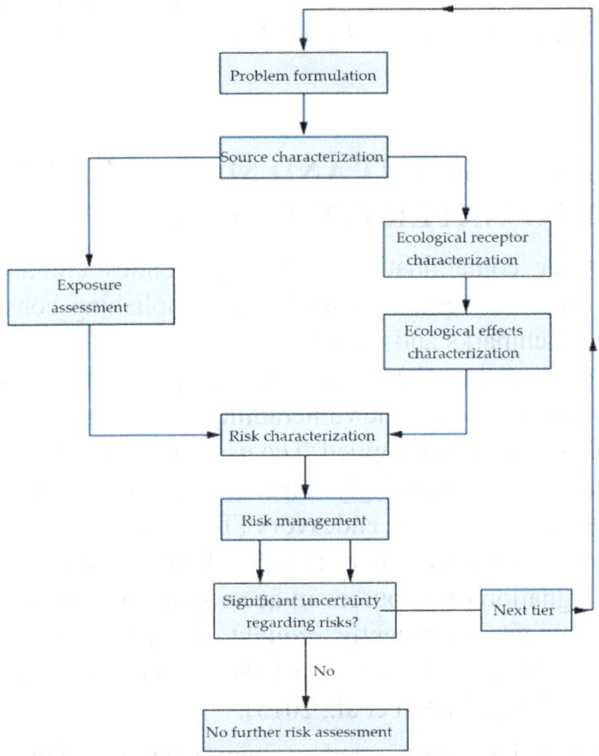

Figure 6.5: Methods for conducting a risk assessment for successive tiers of analysis. Tier 1: Screening level; Tier 2: Quantification of potentially significant risks; Tier 3: Site-specific risk quantification.

[Source: http://www.ilocis.org/documents/chpt55e.htm]

6.10. WATER POLLUTION IN LAKES AND RESERVOIRS

Lakes and supplies accommodate the volumetric stockpiling of watershed inflow and may have long flushing timeframes as contrasted and the fast inflow and surge for a reach in a streaming stream (Serrano and DeLorenzo, 2008; Karmakar et al., 2011). Hence they are of unique worry concerning the maintenance of specific constituents, particularly supplements including types of phosphorus and nitrogen which stimulates eutrophication. Eutrophication is a characteristic maturing process in which the water content turns out to be naturally improved, prompting the control of unfortunate oceanic development, for example, green growth, water hyacinth, etc. (Vartiainen et al., 1995). The eutrophic procedure will in general abatement aquatic life and has inconveniently broken down oxygen impacts. Both regular and social sources of supplements may advance the procedure as represented in Figure 6.6, demonstrating a schematic posting of supplement sources and sinks for Lake Sunapee, in the US State of New Hampshire.

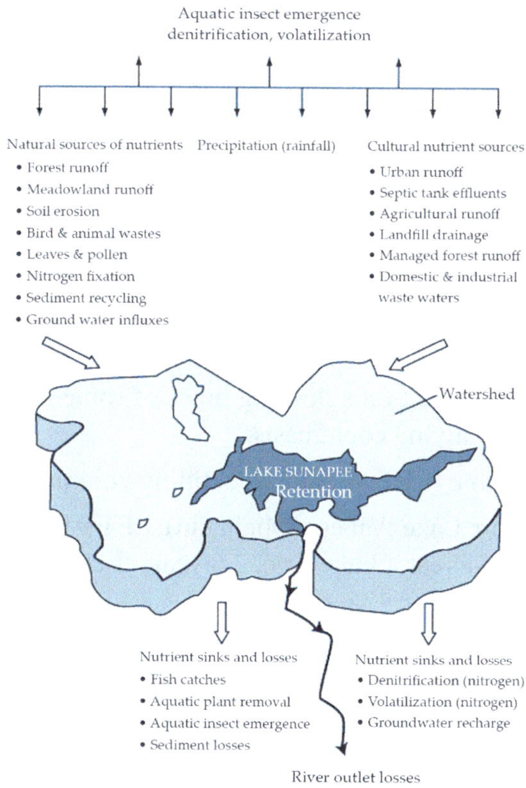

Figure 6.6: Schematic listing of nutrient (nitrogen and phosphorus) sources and sinks for Lake Sunapee, New Hampshire (US).

[Source: http://www.ilocis.org/documents/chpt55e.htm]

Lakes and supplies, obviously, can be inspected and examined to decide their trophic status. Diagnostic examinations, as a rule, start with a fundamental supplement equalization, for example, the accompanying:

(Lake Influent supplements) = (lake profluent supplements) + (supplement maintenance in lake)

This essential parity can be additionally extended to incorporate the different sources that appeared in Figure 6.6.

Flushing time means that the relative maintenance parts of a lake system. Shallow lakes, for example, Lake Erie, have moderately short flushing occasions and are related to cutting edge eutrophication since shallow lakes regularly are progressively helpful for oceanic plant development (Shah et al., 2011; Rahman et al., 2014). Profound lakes, for example, Lake Tahoe and Lake Superior have long flushing periods, which are typically connected with lakes with negligible eutrophication in light of the fact that up to right now, they have not been over-burden and furthermore on the grounds that their extraordinary profundities are not helpful for broad aquatic plant development aside from in the epilimnion (upper zone). Lakes in this classification are commonly delegated oligotrophic, on the basis that they are generally low in supplements and bolster negligible amphibian development, for example, green growth (Chen et al., 2013).

It is important to look at the flushing times of some significant US lakes utilizing the accompanying count basis:

Lake flushing time (LFT) = (lake stockpiling volume)/(lake outpouring)

A few models are Lake Wabesa (Michigan), LFT = 0.30 years; Houghton Lake (Michigan), 1.4 years; Lake Erie, 2.6 years; Lake Superior, 191 years; Lake Tahoe, 700 years.

In spite of the fact that the connection between the procedure of eutrophication and supplement content is unpredictable, phosphorous is commonly perceived as the constraining supplement. In light of completely blended conditions, the specialists have detailed that algal blossoms will, in general, happen if phosphorous surpasses 0.01 mg/l and nitrogen esteems surpass 0.3 mg/l. In stratified lakes and stores, low dissolved oxygen levels in the hypolimnion are early indications of eutrophication. Later PC models are accessible for reenacting nitrogen/phosphorous cycles with temperature varieties.

6.11. WATER POLLUTION IN ESTUARIES

An estuary is a middle of the road path of water between the mouth of a waterway and ocean coast. This path is included a waterway mouth channel reach with stream inflow (freshwater) from upstream and outpouring release on the downstream side into a continually changing tail-water level of ocean water (saltwater). Estuaries are consistently influenced by tidal changes and are among the most intricate waterways experienced in water contamination control (Elderfield and Hepworth, 1975; Nordmyr et al., 2008). The predominant aspects of an estuary are variable saltiness, a salt wedge or interface among salt and new water, and frequently enormous territories of shallow, turbid water overlying mud pads and salt bogs. Supplements are to a great extent provided to an estuary from the in-flowing stream and consolidate with the ocean water living space to give a productive generation of biota and ocean life. Particularly wanted are shellfishes reaped from estuaries (Kwokal et al., 2002; Gowri et al., 2008).

From a water contamination point of view, estuaries are exclusively perplexing and for the most part require unique examinations utilizing broad field studies and PC demonstrating (Magesh et al., 2011).

6.12. WATER POLLUTION IN MARINE ENVIRONMENTS

Seas might be seen as a definitive receiving water or sink since wastes conveyed by streams at long last release into this marine condition (Nendza et al., 1997; Brennecke et al., 2016). Despite the fact that seas are immense assortments of saltwater with apparently boundless osmosis limit, contamination will in general curse coastlines and further influence marine life. Wellsprings of marine poisons incorporate a considerable lot of those experienced in land-based wastewater conditions in addition to additional as identified with marine activities (Lopez-Torres et al., 1988). A constrained rundown is given beneath:

1. Domestic sludge and sewage, shipboard wastes, industrial wastes, solid waste.
2. Fishery squanders, silt and supplements from waterways and land overflow.
3. Oil spills, seaward oil investigation, and generation waste dig activities.

4. Heat, radioactive wastes, chemical wastes, herbicides and pesticides.

Every one of the above requires exceptional taking care of and techniques for control. The release of household sewage and sewage sludge through sea outfalls is maybe the significant wellspring of marine contamination (Bat et al., 1999; Ozmen et al., 2006).

6.13. TECHNIQUES FOR REDUCING POLLUTION IN WASTEWATER DISCHARGES

Huge scale wastewater treatment is normally done by municipalities, sanitary districts, commercial enterprise industries, and different contamination control commissions. The reason here is to portray contemporary techniques for city wastewater treatment and afterward to give a few bits of knowledge with respect to the treatment of industrial wastes and further developed strategies (Sánchez-Avila et al., 2009; Iordache and Dunea, 2013).

All in all, all procedures of wastewater treatment might be gathered into biological, physical or chemical sorts, and at least one of these might be utilized to accomplish an ideal gushing item (Villaescusa-Celaya et al., 2000). This arrangement gathering is generally fitting in the comprehension of wastewater treatment approaches and is classified in Table 6.1.

Table 6.1: The General Classification of Wastewater Treatment Operations and Processes

Physical Operations	Chemical Processes	Biological Processes
Flow measurement Screening/grit removal Mixing Filtration Distillation Freezing Reverse osmosis Flocculation Drying Sedimentation Flotation Centrifuging	Adsorption Ion exchange Electrodialysis Disinfection Chemical oxidation Neutralization Precipitation Chemical reduction Incineration	Aerobic action Aerobic-anaerobic Combinations Anaerobic action

6.14. CONTEMPORARY METHODS OF WASTEWATER TREATMENT

The inclusion here is constrained and is planned to give a theoretical outline of the ebb and flow wastewater treatment rehearses the world over as opposed to point by point structure information. Civil wastewaters alongside some intermixing of commercial/industrial wastes are treated in the system regularly utilizing essential, auxiliary and tertiary treatment as follows:

Primary treatment system: Pre-treat → Primary settling → Disinfection (chlorination) → Effluent

Secondary treatment system: Pre-treat → Primary settling → Biological unit → Second settling → Disinfection (chlorination) → Effluent to stream

Tertiary treatment system: Pre-treat → Primary settling → Biological unit → Second settling → Tertiary unit → Disinfection (chlorination) →Effluent to stream

Figure 6.7 further displays a schematic diagram of a conservative wastewater treatment system. Summary descriptions of the above procedures follow.

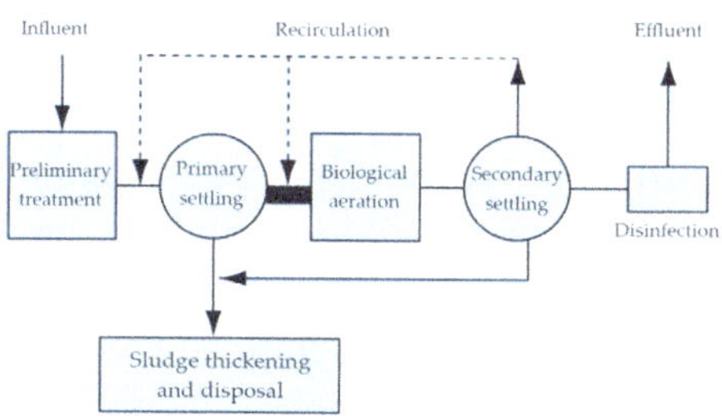

Figure 6.7: Schematic diagram of conventional wastewater treatment.

[Source: http://www.ilocis.org/documents/chpt55e.htm]

6.14.1. Primary Treatment

The fundamental target of essential treatment for metropolitan wastewaters, incorporating household sewage blended with some commercial/industrial wastes, is to evacuate suspended solids and clean the wastewater, to make it appropriate for organic treatment (Lin and Chen, 1997; Gogate and Pandit, 2004). After some pre-treatment taking care of, for example, screening, coarseness expulsion, and comminution, the fundamental procedure of essential sedimentation is the settling of the crude wastewater in enormous settling tanks for periods as long as a few hours. This procedure expels from 50 to 75% of the all-out suspended solids, which are drawn off as a sub-current sludge gathered for isolated treatment. The flood emanating from the procedure at that point is coordinated for auxiliary treatment. In specific cases, synthetic concoctions might be utilized to improve the level of essential treatment (Sonune and Ghate, 2004; Renou et al., 2008).

6.14.2. Secondary Treatment

The part of the natural substance of the wastewater which is finely suspended or broken down and not expelled in the essential procedure is treated by optional treatment. The, for the most part, acknowledged types of optional treatment in like manner use incorporate streaming channels, organic contractors, for example, activated sludge, rotating discs, aerated pond systems, waste stabilization ponds, and land application methods including wetland systems (Comas et al., 2001; Engin and Demir, 2006). These frameworks will be perceived as utilizing natural procedures of some structure or another. The most widely recognized of these procedures are briefly discussed below.

- *Organic contactor systems.* Streaming filters are perhaps the most primary type of this strategy for optional treatment are still generally utilized with some improved techniques for application. In this treatment, the gushing from the essential tanks is applied consistently onto a bed of media, for example, rock or engineered plastic media. Uniform circulation is cultivated commonly by streaming the fluid from punctured funneling pivoted over the bed irregularly or ceaselessly as per the ideal procedure. Contingent upon the pace of natural and hydraulic loadings, streaming channels can evacuate up to 95% of the natural substance, generally dissected as biochemical oxygen demand (BOD). There are various other later organic contactor frameworks being used which can give treatment evacuations in a similar range; a portion of these techniques offer extraordinary points of interest, especially pertinent in certain restricting conditions, for example, space, atmosphere, etc. It is to be noticed that an after auxiliary settling tank is viewed as a vital piece of finishing the procedure. In auxiliary settling, a few so-called humus sludges are drawn off as an undercurrent, and the flood is released as an optional profluent (Karimi et al., 2011).

- *Actuated sludge.* In the most widely recognized type of this organic procedure, essentially treated emanating streams into an activated sludge unit tank containing a formerly existing natural suspension called activated sludge. This blend is alluded to as mixed liquor suspended solids (MLSS) and is given a contact period ordinarily running from a few hours as long as 24 hours or more, contingent upon the ideal outcomes.

During this period the blend is exceptionally circulated and agitated to promote aerobic biological activity. As the procedure finishes, a part of the blend (MLSS) is attracted off and came back to the influent for the continuation of the activation process. Optional settling is given after the activated sludge unit to settle out the activated sludge suspension and releasing an explained flood as a profluent. The procedure is equipped for evacuating up to about 95% of the influent BOD.

6.14.3. Tertiary Treatment

The third degree of treatment might be given where a higher level of pollutant evacuation is required. This type of treatment may commonly incorporate sand filtration, land disposal methods, wetlands and other systems, stabilization ponds that further balance out the auxiliary gushing.

6.14.4. Disinfection of Effluents

Sterilization is usually required to diminish microscopic organisms and pathogens to satisfactory levels. Chlorination, ultraviolet light, ozone and chlorine dioxide are the most generally utilized procedures.

6.14.5. Overall Wastewater Treatment Plant Efficiency

Wastewaters incorporate a wide scope of constituents which for the most part are named suspended and dissolved solids, organic constituents and inorganic. The proficiency of a treatment system can be estimated as far as the rate expulsion of these constituents. Regular parameters of estimation are:

1. BOD: biochemical oxygen demand, measured in mg/l.
2. Nitrogen forms: including ammonia and nitrate, measured in mg/l (nitrate is of specific concern as a nutrient in eutrophication).
3. COD: chemical oxygen demand, measured in mg/l.
4. Coliform bacteria counts: measured as the most possible number

per 100 ml (Escherichia and fecal coliform bacteria are the most common pointers).

5. Phosphate: measured in mg/l (also of specific concern as a nutrient in eutrophication).
6. TSS: total suspended solids, measured in mg/l.
7. Ph: the degree of acidity, measured as a number from 1 (most acid) to 14 (most alkaline).

6.14.6. Industrial Wastewater Treatment

6.14.6.1. Types of Industrial Wastes

Industrial (non-local) wastes are various and shift significantly in the structure; they might be exceptionally acidic or alkaline, and frequently require an itemized research center examination. Specific treatment might be important to render them harmless before release. Poisonous quality is of incredible worry in the disposal of industrial wastewaters.

Common industrial wastes include petroleum, pulp and paper, food processing, tannery, slaughterhouse, cannery, laundry, textile, meat and poultry sugar, oil, hog feeding, rendering, clothing, meat and poultry, and numerous others. The first step in treatment plan development is an industrial waste overview, which gives information on changes in stream and waste qualities. Unfortunate waste attributes can be outlined as follows:

1. Soluble organics causing depletion of dissolved oxygen
2. Colour and turbidity
3. Refractory substances resistant to biodegradation
4. Suspended solids
5. Heavy metals, cyanide, and toxic organics
6. Trace organics
7. Nitrogen and phosphorus

The US EPA has additionally characterized a rundown of lethal natural and inorganic synthetic compounds with explicit limitations in giving release licenses. The rundown incorporates in excess of 100 mixes and is too long to even think about reprinting here, yet might be mentioned from the EPA.

6.14.6.2. Treatment Methods

The treatment of industrial wastes is more sensitive than the treatment of household waste; in any case, where amiable to organic decrease, they are typically treated utilizing strategies like those recently depicted (auxiliary/tertiary natural treatment methods) for metropolitan systems.

Waste stabilization ponds are a typical technique for natural wastewater treatment where the adequate land region is accessible. Course through lakes is commonly arranged by their bacterial movement as aerobic, facultative or anaerobic. Circulated air through lakes is provided with oxygen by diffused or mechanical air circulation systems. At the point when industrial waste in-plant activities and procedures are dissected at their source, they frequently can be controlled in order to forestall noteworthy contaminating releases.

Distribution methods are significant methodologies in contamination avoidance programs. Contextual investigations for reusing plan for a leather tannery wastewater gushing are accessible in writing which incorporates chrome recuperation/reuse alongside the total distribution of all tannery wastewaters with no profluent to any stream aside from in crises. The stream graph for this system is displayed in Figure 6.8.

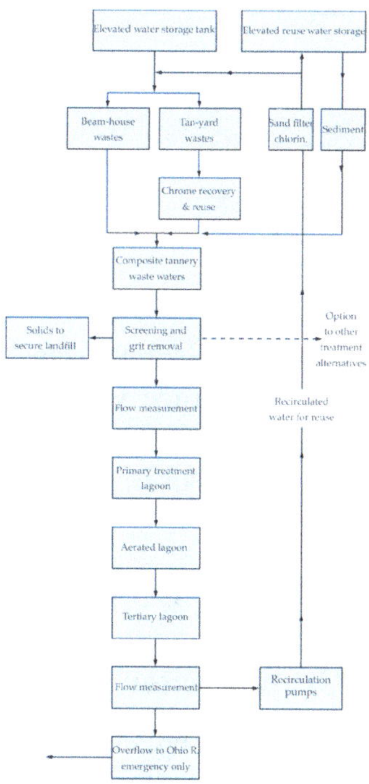

Figure 6.8: Flow diagram for tannery wastewater effluent recycling system.

[Source: http://www.ilocis.org/documents/chpt55e.htm]

6.14.7. Advanced Methods of Wastewater Treatment

A number of innovative approaches are accessible for higher degrees of elimination of contamination constituents as may be essential. A general listing includes:

1. Filtration (sand and multimedia)
2. Wetlands.
3. Ammonia stripping
4. Algae harvesting
5. Reverse osmosis
6. Carbon adsorption
7. Electrodialysis
8. Distillation
9. Chemical precipitation
10. Land application
11. Nitrification
12. Reclamation of effluents
13. Micro-straining
14. Ion exchange
15. Denitrification

The most proper procedure for any circumstance must be resolved based on the quality and amount of the crude wastewater, the receiving water prerequisites and, obviously, costs. For additional reference, which contains a section for advanced wastewater treatment.

6.15. GROUNDWATER POLLUTION CONTROL

6.15.1. Importance of Groundwater

Since the world's water supplies are broadly extricated from springs, it is most significant that these wellsprings of supply be secured. It is evaluated that over 95% of the world's accessible freshwater supply is underground; in the United States around half of the drinking water originates from wells, as per the 1984 US Geological Survey. Since underground water contamination and development are of unobtrusive and inconspicuous nature, less consideration some of the time is given to the examination and control of this type of water degradation than to surface water contamination, which is unmistakably progressively self-evident (Thapinta and Hudak, 2003; Liu et al., 2007).

6.15.2. Sources of Underground Pollution

Figure 6.11 displays the hydrologic cycle with superimposed bases of groundwater pollution. A comprehensive listing of the possible bases of underground contamination is widespread; though, for design the most apparent sources include:

1. Industrial waste discharges
2. Irrigation systems
3. polluted ponds with permeable bottoms
4. Artificial recharge
5. Septic tank tile fields and leaching pits
6. Mining operations
7. Roadway de-icing salts.
8. Solid and hazardous waste disposal
9. Polluted streams in contact with aquifers
10. Seawater encroachment
11. Spills
12. Disposal wells
13. Underground storage tanks such as for petroleum

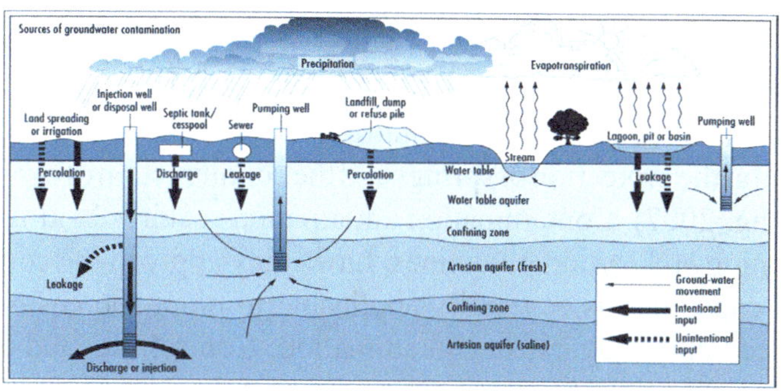

Figure 6.9: Hydrologic cycle and sources of groundwater contamination.

[Source: http://www.ilocis.org/documents/chpt55e.htm]

Particular pollutants in underground pollution are further characterized as:

1. Unwanted chemical constituents inorganic and organic (e.g., potassium, sulfate, chloride, sodium iron, manganese).
2. Total dissolved solids and total hardness.
3. Poisonous constituents: chromium, nitrate, arsenic, phenols, copper, dissolved mercury, lead, cyanide.
4. Unwanted physical features – taste, odor and color.
5. Herbicides and pesticides – chlorinated hydrocarbons and others.
6. Radioactive materials – numerous forms of radioactivity.
7. Biological – bacteria, parasites, viruses and so on.
8. Caustic (high ph) or acid (low ph).

Of the above mentioned, nitrates are of uncommon concern in both groundwaters and surface waters. In groundwater supplies, nitrates can cause the malady methemoglobinaemia (infant cyanosis). They further reason impeding eutrophication impacts in surface waters and happen in a wide scope of water resources.

6.15.3. Pollution Prevention

The favored methodology for the safety of groundwater resources is contamination anticipation. In spite of the fact that drinking water norms by and large apply to the utilization of groundwater supplies, the crude water supplies need protection from tainting. Administrative entities, for example, services of health, common resource organizations, and environmental protection agencies are commonly liable for such exercises (Kumar and Shah, 2006).

Groundwater contamination control endeavors are to a great extent coordinated at the protection of springs and the counteraction of contamination (Yong-Sheng, 2007). Contamination anticipation requires land-use controls through zoning and certain guidelines. Laws may apply to the counteraction of explicit capacities as especially applicable to point sources or activities which conceivably may cause contamination. Control via land-use zoning is a groundwater protection device that is best at the municipal or area level of government. Spring and wellhead protection programs as explained underneath are leading instances of contamination prevention.

An aquifer protection program needs setting up the limits of the aquifer and it's energizing territories. Aquifers might be of an unconfined or restricted kind, and accordingly should be broken down by a hydrologist to make this assurance. Most significant aquifers are commonly notable in developed nations, however, different territories may need field examinations and hydrogeologic investigation. The key component of the program in the protection of the aquifer from water quality debasement is control of land use over the aquifer and its revive territories.

Wellhead protection is an increasingly authoritative and constrained methodology that applies to the energize zone adding to a specific well. The US government by amendments passed in 1986 to the Safe Drinking Water Act (SDWA) (1984) presently necessitates that particular wellhead protection regions be built up for open inventory wells. The wellhead protection area (WHPA) is characterized in the SDWA as "the surface and subsurface zone encompassing a water well or well field, providing an open water supply system, through which contaminants are reasonably possible to push toward and arrive at such water well or well field." The fundamental target in the WHPA program, as laid out by the US EPA (1987), is the depiction of well protection territories dependent on chose criteria, well activity and hydrogeologic contemplation.

REFERENCES

1. Alamu, S. A. (2013). Analysis of seedling subsidy policy and Cocoa production in South-West Nigeria. *Journal of Educational and Social Research, 3*(4), 59.
2. Aliyu, Y. A., Musa, I. J., & Youngu, T. T. (2013). Appraisal of sulfur contaminants from transportation in urban Zaria, Nigeria. *Academic Journal of Interdisciplinary Studies, 2*(10), 155.
3. Bashiri, E., Bashiri, J., & Karimi, F. (2013a). Investigating the recent drought effects on underground water resources in the province of Kurdistan from a quantitative and qualitative point of view (Case study: Dehgolan and Ghorveh plains). *International Letters of Natural Sciences, 3,* 1–55.
4. Bashiri, E., Bashiri, J., & Karimi, F. (2013b). Investigating the recent drought effects on underground water resources in the province of Kurdistan from a quantitative and qualitative point of view (Case study: Dehgolan and Ghorveh plains). *International Letters of Natural Sciences, 3, 1–20.*
5. Bat, L., Gündoğdu, A., Sezgin, M., Çulha, M., Gönlügür, G., & Akbulut, M. (1999). Acute toxicity of zinc, copper and lead to three species of marine organisms from the Sinop Peninsula, Black Sea. *Turkish Journal of Biology, 23*(4), 537–544.
6. Borah, D. K., & Bera, M. (2003). Watershed-scale hydrologic and nonpoint-source pollution models: Review of mathematical bases. *Transactions of the ASAE, 46*(6), 1553.
7. Brawer, J. M. (1999). Antidegradation Policy and Outstanding National Resource Waters in the Northern Rocky Mountain States. *Pub. Land & Resources L. Rev., 20,* 13.
8. Brennecke, D., Duarte, B., Paiva, F., Caçador, I., & Canning-Clode, J. (2016). Microplastics as vector for heavy metal contamination from the marine environment. *Estuarine, Coastal and Shelf Science, 178,* 189–195.
9. Byron, E. R., & Goldman, C. R. (1989). Land-use and water quality in tributary streams of Lake Tahoe, California-Nevada. *Journal of Environmental Quality, 18*(1), 84–88.
10. Chan, C. W., & Huang, G. H. (2003). Artificial intelligence for management and control of pollution minimization and mitigation

processes. *Engineering Applications of Artificial Intelligence, 16*(2), 75–90.

11. Characklis, G. W., & Wiesner, M. R. (1997). Particles, metals, and water quality in runoff from large urban watershed. *Journal of Environmental Engineering, 123*(8), 753–759.
12. Chen, Y., Niu, Z., & Zhang, H. (2013). Eutrophication assessment and management methodology of multiple pollution sources of a landscape lake in North China. *Environmental Science and Pollution Research, 20*(6), 3877–3889.
13. Chilson, J. A. (1998). Keeping Clean Waters Clean: Making the Clean Water Act's Antidegradation Policy Work. *U. Mich. JL Reform, 32*, 545.
14. Chugh, L. C., HANEMANN, M., & Mahapatra, S. (1978). Impact of pollution control regulations on the market risk of securities in the US. *Journal of Economic Studies, 5*(1), 64–70.
15. Chukwu, B. N., Ezebuiro, V. O., Samuel, E. S., & Nwachukwu, K. C. (2013). Gender differential in the incidence of diabetes mellitus among the patients in Udi Local Government Area of Enugu state, Nigeria. *International Letters of Natural Sciences, 4, 1–20.*
16. Chutter, F. M. (1972). An empirical biotic index of the quality of water in South African streams and rivers. *Water Research, 6*(1), 19–30.
17. Comas, J., Dzeroski, S., Gibert, K., R-Roda, I., & Sànchez-Marrè , M. (2001). Knowledge discovery by means of inductive methods in wastewater treatment plant data. AI Communications, 14(1), 45–62.
18. Crandall, C. A., Katz, B. G., & Hirten, J. J. (1999). Hydrochemical evidence for mixing of river water and groundwater during high-flow conditions, lower Suwannee River basin, Florida, USA. Hydrogeology Journal, 7(5), 454–467.
19. Deflandre, A., Williams, R. J., Elorza, F. J., Mira, J., & Boorman, D. B. (2006). Analysis of the QUESTOR water quality model using a Fourier amplitude sensitivity test (FAST) for two UK rivers. Science of the Total Environment, 360(1–3), 290–304.
20. Demirak, A., Yilmaz, F., Tuna, A. L., & Ozdemir, N. (2006). Heavy metals in water, sediment and tissues of Leuciscus cephalus from a stream in southwestern Turkey. *Chemosphere, 63*(9), 1451–1458.

21. Dillon, P. J., & Rigler, F. H. (1974). A test of a simple nutrient budget model predicting the phosphorus concentration in lake water. *Journal of the Fisheries Board of Canada, 31*(11), 1771–1778.
22. Dosskey, M. G. (2001). Toward quantifying water pollution abatement in response to installing buffers on crop land. *Environmental Management, 28*(5), 577–598.
23. Eguabor, V. (1998). Strategies for Teaching water pollution in secondary schools. *STAN Journal Environmental Education Series,* 2, 33–49.
24. Elderfield, H., & Hepworth, A. (1975). Diagenesis, metals and pollution in estuaries. *Marine Pollution Bulletin, 6*(6), 85–87.
25. Ellis, M. M. (1937). Detection and measurement of stream pollution. *Bulletin of the Bureau of Fisheries;(United States), 48*(22), 11–30.
26. Engin, G. O., & Demir, I. (2006). Cost analysis of alternative methods for wastewater handling in small communities. *Journal of Environmental Management, 79*(4), 357–363.
27. Galambos, I., Molina, J. M., Járay, P., Vatai, G., & Bekássy-Molnár, E. (2004). High organic content industrial wastewater treatment by membrane filtration. *Desalination, 162*, 117–120.
28. Gbamanja, S. P. T. (1998). The Pedagogy of water pollution. *STAN Journal "Environmental Education Series," 2, 43.*
29. Gogate, P. R., & Pandit, A. B. (2004). A review of imperative technologies for wastewater treatment II: hybrid methods. *Advances in Environmental Research, 8*(3–4), 553–597.
30. Gordon, S. I. (1985). Water Quality Models. In *Computer Models in Environmental Planning* (Vol. 1, pp. 14–41). Springer, Boston, MA.
31. Gowri, V. S., Ramachandran, S., Ramesh, R., Pramiladevi, I. R. R., & Krishnaveni, K. (2008). Application of GIS in the study of mass transport of pollutants by Adyar and Cooum Rivers in Chennai, Tamilnadu. *Environmental Monitoring and Assessment, 138*(1–3), 41–49.
32. Grema, L. U., Abubakar, A. B., & Obiukwu, O. O. (2013). Carbon emission control measures. *International Letters of Natural Sciences, 3, 1–20.*
33. Griffith, G. E., Omernik, J. M., Wilton, T. F., & Pierson, S. M. (1994). Ecoregions and subregions of Iowa: a framework for water quality assessment and management. *Journal of the Iowa Academy of Science: JIAS, 101*(1), 5–13.

34. Grifoll, M., Del Campo, A., Espino, M., Mader, J., González, M., & Borja, Á. (2013). Water renewal and risk assessment of water pollution in semi-enclosed domains: Application to Bilbao Harbour (Bay of Biscay). *Journal of Marine Systems*, *109*, S241–S251.
35. Hagedorn, C., Robinson, S. L., Filtz, J. R., Grubbs, S. M., Angier, T. A., & Reneau, R. B. (1999). Determining sources of fecal pollution in a rural Virginia watershed with antibiotic resistance patterns in fecal streptococci. *Appl. Environ. Microbiol.*, *65*(12), 5522–5531.
36. Halder, J. N., & Islam, M. N. (2015). Water pollution and its impact on the human health. *Journal of Environment and Human*, *2*(1), 36–46.
37. Harleston, J. (1996). What is Antidegradation Policy: Does Anyone Know. *SC Envtl. LJ*, *5*, 33.
38. Hersh, C. M. (2009). The Clean Water Act's Antidegradation Policy and Its Role in Watershed Protection in Washington State. *Hastings W.-Nw. J. Envt'l L. & Pol'y*, *15*, 217.
39. Hines, N. W. (1966). Nor Any Drop to Drink: Public Regulation of Water Quality Part I: State Pollution Control Programs. *Iowa L. Rev.*, *52*, 186.
40. Huang, G. H., & Xia, J. (2001). Barriers to sustainable water-quality management. *Journal of Environmental Management*, *61*(1), 1–23.
41. Imai, A., Fukushima, T., Matsushige, K., Kim, Y. H., & Choi, K. (2002). Characterization of dissolved organic matter in effluents from wastewater treatment plants. *Water research*, *36*(4), 859–870.
42. Iordache, ş., & dunea, d. (2013). Multivariate analysis applied to physicochemical parameters of treated wastewater effluents discharged in Teleajen River. *Rev. Roum. Chim*, *58*(7–8), 717–723.
43. James, E., & Joyce, M. (2004). Assessment and management of watershed microbial contaminants. *Critical Reviews in Environmental Science and Technology*, *34*(2), 109–139.
44. Kahn, H. D., & Rubin, M. B. (1989). Use of statistical methods in industrial water pollution control regulations in the United States. *Environmental Monitoring and Assessment*, *13*(2–3), 129–148.
45. Karimi, A. R., Mehrdadi, N., Hashemian, S. J., Nabi-Bidhendi, G. R., & Tavakkoli-Moghaddam, R. (2011). Using of the fuzzy topsis and fuzzy ahp methods for wastewater treatment process selection. *International Journal of Academic Research*, *3*(1), 3–36.

46. Karmakar, S., Haque, S. S., Hossain, M. M., & Shafiq, M. (2011). Water quality of Kaptai reservoir in Chittagong Hill Tracts of Bangladesh. *Journal of Forestry Research, 22*(1), 87–92.

47. Karr, J. R., & Chu, E. W. (2000). Introduction: Sustaining living rivers. In *Assessing the Ecological Integrity of Running Waters* (Vol. 1, pp. 1–14). Springer, Dordrecht.

48. Karr, J. R., & Dudley, D. R. (1981). Ecological perspective on water quality goals. *Environmental Management, 5*(1), 55–68.

49. Kaushal, S. S., Groffman, P. M., Band, L. E., Elliott, E. M., Shields, C. A., & Kendall, C. (2011). Tracking nonpoint source nitrogen pollution in human-impacted watersheds. *Environmental Science & Technology, 45*(19), 8225–8232.

50. Kaze, P. D., & Gam, K. P. (2013). Survey of Cysticercus (Bladder Worm) in meat sold for consumption in Bukuru, Plateau State Nigeria. *Academic Journal of Interdisciplinary Studies, 2*(12), 45.

51. Kimes, J. K. (1995). The regulation of concentrate disposal in Florida. *Desalination, 102*(1–3), 87–92.

52. Kumar, M. D., & Shah, T. (2006). Groundwater pollution and contamination in India: the emerging challenge. *IWMI-TATA Water Policy Program Draft Paper, 1*, 14.

53. Kwokal, Ž., Frančišković-Bilinski, S., Bilinski, H., & Branica, M. (2002). A comparison of anthropogenic mercury pollution in Kaštela Bay (Croatia) with pristine estuaries in Öre (Sweden) and Krka (Croatia). *Marine Pollution Bulletin, 44*(10), 1152–1157.

54. Leenheer, J. A. (1981). Comprehensive approach to preparative isolation and fractionation of dissolved organic carbon from natural waters and wastewaters. *Environmental Science & Technology, 15*(5), 578–587.

55. Lewis, K. A., Tzilivakis, J., Warner, D. J., & Green, A. (2016). An international database for pesticide risk assessments and management. *Human and Ecological Risk Assessment: An International Journal, 22*(4), 1050–1064.

56. Liding, C., & Bojie, F. (2000). Farm ecosystem management and control of nonpoint source pollution. *Chinese Journal of Environmental Science, 21*(2), 98–100.

57. Lin, S. H., & Chen, M. L. (1997). Treatment of textile wastewater by chemical methods for reuse. *Water Research, 31*(4), 868–876.

58. Lindegaard, C. (1995). Classification of water-bodies and pollution. In *The Chironomidae* (Vol. 1, pp. 385–404). Springer, Dordrecht.

59. Liquete, C., Udias, A., Conte, G., Grizzetti, B., & Masi, F. (2016). Integrated valuation of a nature-based solution for water pollution control. Highlighting hidden benefits. *Ecosystem Services*, *22*, 392–401.

60. Liu, S. Q., Li, J. G., Mei, M., & Dong, D. L. (2007). Groundwater pollution from underground coal gasification. *Journal of china University of Mining and Technology*, *17*(4), 467–472.

61. Loos, R., Gawlik, B. M., Locoro, G., Rimaviciute, E., Contini, S., & Bidoglio, G. (2009). EU-wide survey of polar organic persistent pollutants in European river waters. *Environmental pollution*, *157*(2), 561–568.

62. Lopez-Torres, A. J., Prieto, L., & Hazen, T. C. (1988). Comparison of the in situ survival and activity ofKlebsiella pneumoniae andEscherichia coli in tropical marine environments. *Microbial Ecology*, *15*(1), 41–57.

63. Loya, Y. (1975). Possible effects of water pollution on the community structure of Red Sea corals. *Marine Biology*, *29*(2), 177–185.

64. Ma, H., Allen, H. E., & Yin, Y. (2001). Characterization of isolated fractions of dissolved organic matter from natural waters and a wastewater effluent. *Water Research*, *35*(4), 985–996.

65. Magesh, N. S., Chandrasekar, N., & Roy, D. V. (2011). Spatial analysis of trace element contamination in sediments of Tamiraparani estuary, southeast coast of India. *Estuarine, Coastal and Shelf Science*, *92*(4), 618–628.

66. Mitch, W. A., Gerecke, A. C., & Sedlak, D. L. (2003). A N-nitrosodimethylamine (NDMA) precursor analysis for chlorination of water and wastewater. *Water research*, *37*(15), 3733–3741.

67. Musa, H. (2013). Water Pollution. "*Water Pollution in Focus" Contribution from the Former Suleja LGA Chairman*, *21*(6), 9–39.

68. Najm, I., & Trussell, R. R. (2001). NDMA formation in water and wastewater. *Journal-American Water Works Association*, *93*(2), 92–99.

69. Nendza, M., Herbst, T., Kussatz, C., & Gies, A. (1997). Potential for secondary poisoning and biomagnification in marine organisms. *Chemosphere*, *35*(9), 1875–1885.

70. Nordmyr, L., Åström, M., & Peltola, P. (2008). Metal pollution of estuarine sediments caused by leaching of acid sulphate soils.

Estuarine, Coastal and Shelf Science, *76*(1), 141–152.

71. Novotny, V. (1994). *Water Quality: Prevention, Identification and Management of Diffuse Pollution* (Vol. 1, pp. 1–40). Van Nostrand-Reinhold Publishers.
72. Nwona, H. A. (2013). Climate change: causes, effects and the need for science education for sustainable development. *Mediterranean Journal of Social Sciences*, *4*(8), 35.
73. Nzewi, E. U., Gray, D. D., & Houck, M. H. (1985). Optimal design program for gravity sanitary sewers. *Civil Engineering Systems*, *2*(3), 132–141.
74. Obilonu, A. N., Chijioke, C., Igwegbe, W. E., Ibearugbulem, O. I., & Abubakar, Y. F. (2013). Water quality challenges and impact. *International Letters of Natural Sciences, 4, 1–70.*
75. Ogwuazor, K. E. (1998). Teaching water pollution by guided discovery. *STAN Journal Environmental Education Series (2), 52.*
76. Oladunjoye, H. T., Olasunkanmi, N., & Olaleye, A. O. (2013). Geophysical evaluation of magnetic data of Okenugbo area, Age-Iwoye, Southwestern Nigeria. *International Letters of Natural Sciences, 3, 1–30.*
77. Olaniran, N. S. (1995). Environment and health: an introduction. *Environment and Health. Lagos. Micmillan Nig. Pub. Co for NCF*, 1, 34–151.
78. Omernik, J. M., & Griffith, G. E. (1991). Ecological regions versus hydrologic units: frameworks for managing water quality. *Journal of Soil and Water Conservation*, *46*(5), 334–340.
79. Owa, F. D. (2013). Water pollution: sources, effects, control and management. *Mediterranean Journal of Social Sciences*, *4*(8), 65.
80. Ozmen, M., Güngördü, A., Kucukbay, F. Z., & Güler, R. E. (2006). Monitoring the effects of water pollution on Cyprinus carpio in Karakaya Dam Lake, Turkey. *Ecotoxicology*, *15*(2), 157–169.
81. Palange, R. C., & Zavala, A. (1987). *Water Pollution Control: Guidelines for Project Planning and Financing* (Vol. 23, pp. 11–33). Washington, DC: World Bank.
82. Puckett, L. J. (1995). Identifying the major sources of nutrient water pollution. *Environmental Science & Technology*, *29*(9), 408A-414A.
83. Rahman, M. M., Bashar, M. A., Farhana, Z., & Hossain, M. Y. (2014). Temporal variation of physicochemical parameters in Kaptai lake,

Bangladesh. *World Journal of Fish and Marine Sciences*, *6*(5), 475–478.

84. Rajinikanth, R., Ganesh, R., Escudie, R., Mehrotra, I., Kumar, P., Thanikal, J. V., & Torrijos, M. (2009). High rate anaerobic filter with floating supports for the treatment of effluents from small-scale agro-food industries. *Desalination and Water Treatment*, *4*(1–3), 183–190.
85. Rajovic, G., & Bulatovic, J. (2013). Geographical view on energetic sources of climate northeastern Montenegro. *International Letters of Natural Sciences, 3, 1–20.*
86. Rankin, E. T. (1995). Habitat indices in water resource quality assessments. *Biological Assessment and Criteria: Tools for Water Resource Planning and Decision Making, 1995*, 181–208.
87. Rathbun, R. E., & Tai, D. Y. (1981). Technique for determining the volatilization coefficients of priority pollutants in streams. *Water Research*, *15*(2), 243–250.
88. Renou, S., Thomas, J. S., Aoustin, E., & Pons, M. N. (2008). Influence of impact assessment methods in wastewater treatment LCA. *Journal of Cleaner Production*, *16*(10), 1098–1105.
89. Ringquist, E. J. (1994). Policy influence and policy responsiveness in state pollution control. *Policy Studies Journal*, *22*(1), 25–43.
90. Safe, S. H. (1998). Hazard and risk assessment of chemical mixtures using the toxic equivalency factor approach. *Environmental Health Perspectives*, *106*(suppl 4), 1051–1058.
91. Sánchez-Avila, J., Bonet, J., Velasco, G., & Lacorte, S. (2009). Determination and occurrence of phthalates, alkylphenols, bisphenol A, PBDEs, PCBs and PAHs in an industrial sewage grid discharging to a Municipal Wastewater Treatment Plant. *Science of the Total Environment*, *407*(13), 4157–4167.
92. Serrano, L., & DeLorenzo, M. E. (2008). Water quality and restoration in a coastal subdivision stormwater pond. *Journal of Environmental Management*, *88*(1), 43–52.
93. Shah, R. D. T., Shah, D. N., & Nesemann, H. (2011). Development of a macroinvertebrate-based Nepal Lake Biotic Index(NLBI): an applied method for assessing the ecological quality of lakes and reservoirs in Nepal. *International Journal of Hydrology Science and Technology*, *1*(1–2), 125–146.

94. Simeon, P. O., & Ambah, B. (2013). Effect of municipal solid waste on the growth of maize (Zea mays L.). *International Letters of Natural Sciences*, *2*, 3–16.

95. Sliva, L., & Williams, D. D. (2001). Buffer zone versus whole catchment approaches to studying land use impact on river water quality. *Water Research*, *35*(14), 3462–3472.

96. Smith, V. K., Desvousges, W. H., & McGivney, M. P. (1983). Estimating water quality benefits: an econometric analysis. *Southern Economic Journal*, 1, 422–437.

97. Sonune, A., & Ghate, R. (2004). Developments in wastewater treatment methods. *Desalination*, *167*, 55–63.

98. Stahl Jr, R. G. (1991). The genetic toxicology of organic compounds in natural waters and wastewaters. *Ecotoxicology and Environmental Safety*, *22*(1), 94–125.

99. Stevens, J. B. (1966). Recreation benefits from water pollution control. *Water Resources Research*, *2*(2), 167–182.

100. Swihart, T., Walton Jr, O. E., & Pennington, M. G. (1986). An antidegradation policy for preserving surface water quality in Florida 1. *JAWRA Journal of the American Water Resources Association*, *22*(4), 665–671.

101. Tamminga, S. (1992). Nutrition management of dairy cows as a contribution to pollution control. *Journal of Dairy Science*, *75*(1), 345–357.

102. Tchounwou, P. B., Centeno, J. A., & Patlolla, A. K. (2004). Arsenic toxicity, mutagenesis, and carcinogenesis–a health risk assessment and management approach. *Molecular and Cellular Biochemistry*, *255*(1–2), 47–55.

103. Thapinta, A., & Hudak, P. F. (2003). Use of geographic information systems for assessing groundwater pollution potential by pesticides in Central Thailand. *Environment International*, *29*(1), 87–93.

104. Tim, U. S., & Jolly, R. (1994). Evaluating agricultural nonpoint-source pollution using integrated geographic information systems and hydrologic/water quality model. *Journal of Environmental Quality*, *23*(1), 25–35.

105. Vartiainen, T., Lampi, P., Tolonen, K., & Tuomisto, J. (1995). Polychlorodibenzo-p-dioxin and polychlorodibenzofuran concentrations in lake sediments and fish after a ground water pollution with chlorophenols. *Chemosphere*, *30*(8), 1439–1451.

106. Villaescusa-Celaya, J. A., Gutiérrez-Galindo, E. A., & Flores-Muñoz, G. (2000). Heavy metals in the fine fraction of coastal sediments from Baja California (Mexico) and California (USA). *Environmental Pollution*, *108*(3), 453–462.

107. Volk, M., Hirschfeld, J., Dehnhardt, A., Schmidt, G., Bohn, C., Liersch, S., & Gassman, P. W. (2008). Integrated ecological-economic modelling of water pollution abatement management options in the Upper Ems River Basin. *Ecological Economics*, *66*(1), 66–76.

108. Wilhm, J. L., & Dorris, T. C. (1968). Biological parameters for water quality criteria. *Bioscience*, 1, 477–481.

109. Wood, A. (1995). Constructed wetlands in water pollution control: fundamentals to their understanding. *Water Science and Technology*, *32*(3), 21–29.

110. Wu, D., Li, W., & Xiao, R. M. (2005). Application of Water Environment Capacity and Total Pollution Quantity Control to Formulation of Effluent Criteria [J]. *Environmental Science and Technology*, *2*, 5–13.

111. Yang, Y. S., & Wang, L. (2010). A review of modelling tools for implementation of the EU water framework directive in handling diffuse water pollution. *Water Resources Management*, *24*(9), 1819–1843.

112. Yoder, C. O., & Rankin, E. T. (1998). The role of biological indicators in a state water quality management process. *Environmental Monitoring and Assessment*, *51*(1–2), 61–88.

113. Yoder, C. O., Miltner, R. J., & White, D. (1999). Assessing the status of aquatic life designated uses in urban and suburban watersheds. In *Proc. Natl. Conf. Retrofit Opportunities for Water Resour. Prot. Urban Environ* (Vol. 1, pp. 16–28).

114. Yong-Sheng, Z. H. A. O. (2007). Groundwater pollution control and remediation. *Journal of Jilin University: Earth Science Edition*, *37*(2), 303–304.

115. Young, R. A., Onstad, C. A., Bosch, D. D., & Anderson, W. P. (1989). AGNPS: A nonpoint-source pollution model for evaluating agricultural watersheds. *Journal of Soil and Water Conservation*, *44*(2), 168–173.

CHAPTER 7

Sustainable and Eco-Friendly Civil Architectures

CONTENTS

7.1. INTRODUCTION

Construction materials are the only unit cell of a structure and in this manner are the basic piece of development systems. It encases the spaces in a structure or goes about as an envelope for spaces in engineering. In the present situation, practical engineering is at its speed. The maintainable design does not just mean to grow trees or utilizing vernacular substances, yet it likewise intends to utilize environmental-friendly structure materials since construction materials are the first and most significant pieces of a structure. Life of a structure is subject to the material being utilized in that building (Guy and Farmer, 2001; Guy and Moore, 2007). Subsequently, the determination of an environmental-friendly material is a quick advance towards the economic and environmental-friendly created surroundings. What material could be environmental-friendly? Saying materials that outcome in a progressively reasonable and moderate development following the solace models required today can be named as environmental-friendly substances (Bradley et al., 2008; Park and Kim, 2018). Along these lines, rather making condition maintainable we could utilize materials that are practical, reasonable, low energy utilization and can be recycled or reused effectively (Mahdavinejad et al., 2013; Garcia et al., 2014). This section manages how green structure materials can sustain our surroundings by diminishing its effect on our mother nature, that could decrease the measure of natural debasement, have less energy consumption, can be re-utilized in a structure itself and the correlation between other man-made materials and their vitality utilization that are being used currently (Oglu Huseynov, 2011; Ragheb et al., 2016).

7.2. SUSTAINABLE CONSTRUCTION MATERIALS

Engineered architectures last longer than chronicles of men and furthermore, through design, planners lead lives past their selves. Architects develop structures as well as they are the people who make a society that goes identified with our surroundings in numerous terms. Architectures may respond to our surroundings either negatively or positively. We, for the most part, utilize term manageable design to make an eco-friendly architecture. In economical engineering, our present activities and choices prompt people in the future. It decreases the negative effect of working in different manners, for example, utilizing present-day materials that create less waste after destruction or rather utilizing waste materials, low energy consuming materials and advancement of spaces in a precise manner (Castro-Lacouture et al., 2009; Florez & Castro-Lacouture, 2013).

Today numerous individuals that are constructing or rebuilding their homes decide to utilize environmental-friendly structure materials. An environmental-friendly structure material is one that expands the productivity of energy utilized and decreases sway on human prosperity and the earth. There is a wide range of materials that can be utilized that are environmental-friendly; from construction to protection, to inside and outside wall, ground surface, and ledge materials (Stadel et al., 2011).

7.3. PARAMETERS FOR MAINTAINABLE CONSTRUCTION MATERIALS

A green structure lessens the general effect of the created environment on human well-being and the indigenous habitat in this way a feasible structure material ought to be environmental-friendly with a long life (Dixit et al., 2012; Estokova and Porhincak, 2015). A material that has the accompanying properties can be called reasonable structure material or green material (Mateus et al., 2013):

i. A recyclable material.
ii. Long-lasting buildings of these materials must be durable so that they will not burn, rot or rust easily.
iii. Effortlessly accessible with the easiness of transport.
iv. Low VOC generating material that will produce fewer toxins for good inside air quality.
v. Must have better thermal mass.
vi. Construction should save and preserve soil.
vii. Enhancement of maintenance and operation.

7.4. CATEGORIZATION OF STRUCTURE MATERIALS

Different structure materials are demonstrated below:

1. Chemical additives and Waterproofing
2. Civil materials
3. Paints, whitewashing, colors wood finishes, and distempering

4. Paving, dado, flooring, and similar finishes
5. Ceiling and roofing
6. Woodwork
7. Sanitary fittings and water supply
8. Windows and doors and
9. Fire-fighting systems
10. Road works
11. Electrical works
12. Excavation work
13. Miscellaneous

7.5. NEED FOR ENVIRONMENTAL-FRIENDLY MATERIALS

Following factors demonstrate the necessity of environmental-friendly materials (Mirjalili et al., 2011):

i. Amazing development in the construction business that relies on limited resources.
ii. Production of construction materials prompts irreversible natural effects.
iii. Utilizing green materials is the most ideal approach to construct a green structure.

Figure 7.1: Stone extraction phenomenon.

[Source: https://theconstructor.org/building/eco-friendly-building-materials/720/]

Stone quarrying prompts eroded slopes, similar to this image demonstrating the area of makrana marble quarry, brick ovens in the edges of the city lead to the covering of topsoil, digging for sand harm the waterway biodiversity and so forth (Khan et al., 2015; Feng et al., 2016).

7.6. DEFINITION OF ENVIRONMENTAL-FRIENDLY MATERIAL

Word reference: depicts an item that has been intended to do the least conceivable harm to the earth. US EPA – EPP program characterizes as: "items or services that have a lesser or decreased impact on human well-being and nature when contrasted against contending items or services that fill a similar need." Along with this, we discuss two issues; one that it does the minimum conceivable natural harm, and two that it is a relative scale as there are not many totally environmental-friendly materials (Fu et al., 2012; Rebah and Siddeeg, 2017). One more point to note when we talk of environmental-friendly development is that it comprises of two sections – Material and Technique. The material can be environmental-friendly itself, for example, Bamboo.

Or on the other hand, regular materials can become environmental-friendly dependent on the construction method that is utilized. For example, the rat-trap bond created by Lauri Baker, which require less number of blocks and are more warmth protecting than ordinary walls and subsequently environmental-friendly (Ukoba et al., 2011).

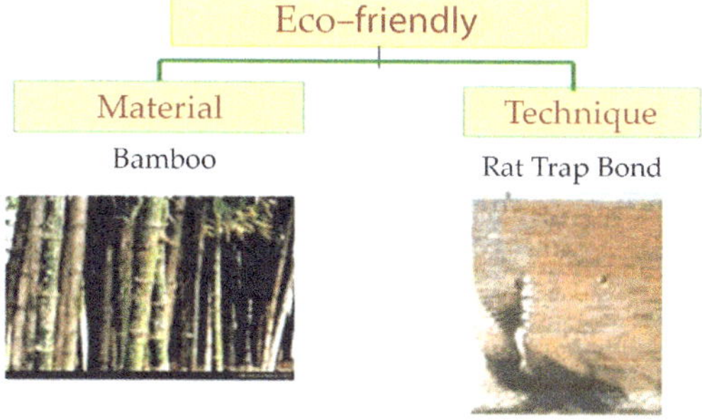

Figure 7.2: Material and operation techniques for environmental-friendly production.

[Source: https://theconstructor.org/building/eco-friendly-building-materials/720/]

7.7. PROPERTIES OF ENVIRONMENTAL-FRIENDLY CONSTRUCTION MATERIALS

If they can help in reducing the energy use in construction during maintenance and operation, materials can be environmental-friendly as well (Jang et al., 2015; Huynh et al., 2018).

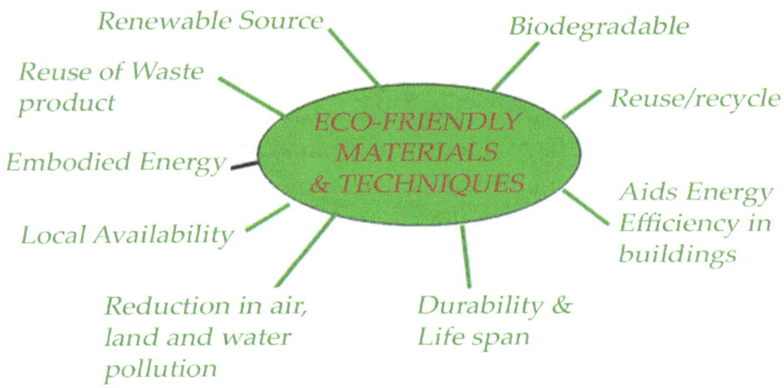

Figure 7.3: Numerous types of environmental-friendly techniques and materials.

[Source: https://theconstructor.org/building/eco-friendly-building-materials/720/]

It is hard to get a substance that has all such assets, and it, therefore, becomes a relative analysis to classify environmental-friendly materials (Asokan et al., 2012).

7.8. SOURCES OF ENVIRONMENTAL-FRIENDLY MATERIAL

7.8.1. Renewable source

Wood from forests that are certified is included in fast renewable sources (Vijayaraghavan & Vincent, 2012; Lopes et al., 2013).

Figure 7.4: Woods as a source of reusable material for building.

[Source: https://theconstructor.org/building/eco-friendly-building-materials/720/]

7.8.2. Salvage of Waste

- Recovered products contain door frames, old plumbing, etc.
- Reprocessed contents contain industrial/agriculture waste, e.g., Bagasse Board.

7.9. EMBODIED ENERGY

The scalar aggregate of energy insertion needed to deliver the item including shipping them to the structure site. Steel and aluminum have the most exemplified energy in light of the fact that to the high energy needed to generate them. Contrasted with timber that needs extremely less energy for creation.

7.10. REDUCED POLLUTION

The utilization of inexhaustible and supportable options bring about diminished contamination. The rundown of contamination types is given beneath (Raja et al., 2013; Wang et al., 2017).

7.10.1. Air Pollution

The use of fewer VOC emanation materials can reduce air pollution, for example, Cement Paints.

Figure 7.5: Eco-friendly paints.

[Source: https://theconstructor.org/building/eco-friendly-building-materials/720/]

7.10.2. Water Pollution

Using materials that avoid leaching can reduce air contamination.

Figure 7.6*:* Materials responsible for reduced leaching.

[Source: https://theconstructor.org/building/eco-friendly-building-materials/720/]

7.10.3. Land Pollution

Air contamination is decreased by utilizing Materials that reuse squander that would some way or another have brought about landfill, for example, Flyash Bricks.

7.11. IMPROVED PERFORMANCE

The performance of common structures can be enhanced by utilizing environmental-friendly materials. The efficiency markers incorporate (Singh et al., 2011; Sekar, 2016).

7.11.1. Reduce Material Consumption

These are vitality effective and furthermore help decrease the load of a structure, for example, Ferrocement.

7.11.2. Durability and Life Span

Practical material that is incredibly tough, or requires low upkeep, for example, PVC pipes.

These materials can be environmental-friendly dependent on how they behave. Utilization of certain material or procedures can decrease the measure of material needed. The more extended the life of a material the lesser it is needed to supplant and in this manner decrease the amount required to create.

7.11.3. Energy Conservation

Most of the inexhaustible materials need less energy during development for example precast sections.

7.11.4. Recyclable

Most of the inexhaustible materials are salvageable or reprocessable as various items, for example, steel, aluminum.

7.11.5. Biodegradable

The vast majority of the inexhaustible materials are decomposable that decay effectively, for example, wood or minerals.

7.12. KINDS OF ENVIRONMENTAL-FRIENDLY CONSTRUCTION MATERIALS

7.12.1. Conventional Environmental-Friendly Materials

1. Sun-dried bricks
2. Bamboo, bamboo matting, bamboo-based plyboard and particleboard
3. Calcined wall panels of phospho-gypsum
4. Pre-formed concrete blocks of cement, slab, lintels, non-structural and structural integrated elements
5. Lightweight cellular concrete blocks
6. Calcium silicate tiles and boards
7. Clay roofing
8. Cement paint
9. Flooring, sealants, Epoxy Resin System, admixtures, and adhesives
10. Water, acrylic-based, and polyurethane chemicals for oxidization removal, rust inhibition, waterproofing
11. Ferro-cement roofing channels
12. Ferro-cement boards for window and door shutters
13. Gypsum board, plaster, blocks, tiles, gypsum plaster fiber, and glass fiber composites
14. Paver blocks and fly-debris sand-lime bricks
15. Marble mosaic tiles
16. Plastic-coated wood plastic components
17. Micro concrete roofing
18. Portland pozzolana cement fly-ash/calcined clay based
19. Moldings and MDF boards
20. Polymerized waterproof compound
21. Portland slag cement
22. Particleboards
23. Ready-mix cement concrete
24. RCC door frames

25. Stone dust
26. Rubber wood finger joint board
27. Waterproof compound, polymer, adhesive, powder

7.12.2. Potential Environmental-Friendly Materials and Methods

1. Bricks from coal washery reject – CBRI, Roorkee
2. Bagasse board – BMTPC
3. Scorched clay fly-debris bricks – CBRI, Roorkee
4. Building slabs from excavation waste – SERC
5. Compacted earth blocks – BMTPC
6. Coir cement board – CBRI, Roorkee
7. Fiber fly-debris cement boards – BMTPC
8. EPS combinations and door shutters – CBRI, Roorkee
9. Rubbery gypsum plasterboards – CBRI, Roorkee
10. Fiber-reinforced concrete ready-made elements, blocks, wall panels, manhole covers – SERC
11. Fly-debris lime cellular concrete – CBRI, Roorkee
12. Fly-debris cement brick, fly-debris cellular concrete, slabs – BMTPC
13. Insulating blocks from rice husk-ash –Central Glass and Ceramic Research Institute, Kolkata
14. Fly-debris lime gypsum brick – BMTPC
15. Non-erodible mud plaster – CBRI, Roorkee
16. Jute fiber polyester – BMTPC
17. Timber from trees such as rubber, poplar, eucalyptus – BMTPC
18. Polytiles – CBRI, Roorkee
19. Prefab brick panel system – CBRI, Roorkee
20. Ready-made walling roofing components – CBRI, Roorkee

7.13. RECOMMENDED ENVIRONMENTAL-FRIENDLY ALTERNATIVES

7.13.1. Structural System

i. Present-day basic frameworks are choices to Cement Concrete (plain/fortified) – concrete, aggregate, sand, steel (Shabbir et al., 2016). Besides, they are base Materials for Steel Systems and R.C.C.

ii. Pozzolana Material substance (Fly-debris/Slag/Calcinated Clay) accomplished through the utilization of Blended Portland Cement (BPC) and/or direct introduction of pozzolana material

iii. Sand and accumulations from pounded trash and/or sintered fly-debris for cement and mortar

Alternative systems:

iv. Recycled steel structures and support bars

v. Ferroconcrete and ready-made segments for columns, pillars, pieces, lofts, overhangs, rooftops and so on.

vi. Ready-Mix Concrete

vii. Use Resinous relieving agents

7.13.2. Stonework

Alternatives to Fired clay bricks, cement concrete blocks, stone, etc. These include:

i. Powdered debris + cement blocks/bricks

ii. Use of Fly-debris + sand + lime blocks/bricks

iii. Aired lightweight BPC concrete slabs

iv. Manufacturing waste-based blocks/bricks

v. Lato blocks (laterite + cement)

vi. Phospho-Gypsum based slabs

7.13.3. Miscellaneous Substitutions

Faswall is an extraordinary option in contrast to customary construction materials. It utilizes squander materials from wood blended in with concrete to make a solid and sturdy construction material. Faswall can be utilized rather than soot squares and is astounding for base construction and wall development. It is normally tough, fire, wind and sound safe, and has brilliant protecting elements.

Two extraordinary decisions for protection are straw parcel and cotton. The straw bundle is utilized for some reasons; it is commonly more affordable than different types of protection and has extraordinary protecting components. It is truly tough and can without much of a stretch help a great deal of weight; extra backings are not normally required, which eliminates the utilization of treated timber (Yun et al., 2010; Choi, 2012).

Cotton protection is a decent option in contrast to fiberglass protection. This sort of protection is gotten from characteristic cotton filaments that are reused into protection. It is extremely helpful and arrives in a comparative batting shape as customary protection. It is likewise simple and safe to introduce and has a high fire rating and brilliant sound protecting characteristics.

Earth mortar and milk paints are superb environmental-friendly decisions for building materials. Earth mortar is made out of the mud and can be utilized as an inside or outside completion. You can control the mortar to make different finishes; including bends, points, and clean completed walls. It should not be painted, as it arrives in an assortment of hues that are normal to the mud.

Milk paints are comprised of milk earth and shades from the earth in an assortment of hues. Milk paints accommodate a wide assortment of hues without the utilization of cruel synthetic compounds that are found in customary house paints.

There are some extraordinary decisions for environmental-friendly ground surface; stopper deck, bamboo, and reused wood flooring are generally phenomenal choices. Stopper flooring is an extraordinary option in contrast to wood; it is ecologically amicable; it just uses the bark from the tree and does not kill the tree thus from harvest. The bark is an extraordinary sustainable asset as it recovers back totally inside nine years. Stopper flooring is entirely agreeable, sturdy, and protecting, and can keep going for a long time.

Bamboo is another extraordinary decision for an inexhaustible asset. Bamboo is not a wood by any means, but instead a grass that lone takes three years to completely develop. At the point when handled, bamboo can have an alluring appearance like wood flooring. It is a solid and strong sort of ground surface and performs particularly like a conventional hardwood floor.

On the off chance that you are resolved to have wood floors, reused wood flooring is an extraordinary choice. This sort of ground surface regularly utilizes hardwood that would some way or another never be utilized again. The wood might be taken from destruction locales or trees that were torn down however never utilized. It is frequently less expensive to plan reused wood rather than a new wood item and is similarly as delightful and tough as new wood.

There are even materials for ledges that are environmental-friendly. Two incredible decisions are reused glass/cast concrete and reused plastic. Reused glass/cast solid uses both reused glass and cement with reused materials in it. The glass utilized in this material cannot be separated any further, so it is a solid match for ledge development. It can likewise be utilized in the washroom and encompassing chimneys.

Reused plastic is another acceptable choice for ledges. You can get various styles utilizing various kinds of reused plastic. It can likewise be utilized in restrooms and different places around the home. It is truly sturdy and can rise to warmth, contingent upon how it is produced and is very water safe.

There is a wide range of environmental-friendly structure materials accessible in the market today. You can, for all intents and purposes, manufacture and outfit a whole home from just environmental-friendly materials. Environmental-friendly materials are useful for the Earth, yet are similarly useful for your wallet. A large number of structural materials that are environmental-friendly have extraordinary protecting elements or are less expensive to create than customary materials.

REFERENCES

1. Asokan, P., Firdoous, M., & Sonal, W. (2012). Properties and potential of bio fibres, bio binders, and bio composites. *Rev. Adv. Mater. Sci, 30*, 254–261.
2. Bradley, K., Gunnarsson-Östling, U., & Isaksson, K. (2008). Exploring environmental justice in Sweden: How to improve planning for environmental sustainability and social equity in an "eco-friendly" context. *Projections, 8*, 68–81.
3. Castro-Lacouture, D., Sefair, J. A., Flórez, L., & Medaglia, A. L. (2009). Optimization model for the selection of materials using a LEED-based green building rating system in Colombia. *Building and Environment, 44*(6), 1162–1170.
4. Choi, N. Y. (2012). Customer Characteristics about Eco-Friendly Apparel Products between'Environmentally Friendly Group'and'Non-environmentally Friendly Group': A Survey of College Students in Daejeon. *Fashion & Textile Research Journal, 14*(5), 791–802.
5. Dixit, M. K., Fernández-Solís, J. L., Lavy, S., & Culp, C. H. (2012). Need for an embodied energy measurement protocol for buildings: A review paper. *Renewable and Sustainable Energy Reviews, 16*(6), 3730–3743.
6. Estokova, A., & Porhincak, M. (2015). Environmental analysis of two building material alternatives in structures with the aim of sustainable construction. *Clean Technologies and Environmental Policy, 17*(1), 75–83.
7. Feng, Y., Bi, Y., Zhao, W., & Zhang, T. (2016). Anionic metal–organic frameworks lead the way to eco-friendly high-energy-density materials. *Journal of Materials Chemistry A, 4*(20), 7596–7600.
8. Florez, L., & Castro-Lacouture, D. (2013). Optimization model for sustainable materials selection using objective and subjective factors. *Materials & Design, 46*, 310–321.
9. Fu, Z., Suo, B., Yun, R., Lu, Y., Wang, H., Qi, S., ... & Matejka, V. (2012). Development of eco-friendly brake friction composites containing flax fibers. *Journal of Reinforced Plastics and Composites, 31*(10), 681–689.
10. Garcia, D. A., Cumo, F., Giustini, F., Pennacchia, E., & Fogheri, A. M. (2014). Eco-architecture and sustainable mobility: An integrated

approach in Ladispoli town. *WIT Transactions on the Built Environment, 142*, 59–68.

11. Guy, S., & Farmer, G. (2001). Reinterpreting sustainable architecture: the place of technology. *Journal of Architectural Education, 54*(3), 140–148.
12. Guy, S., & Moore, S. A. (2007). Sustainable architecture and the pluralist imagination. *Journal of Architectural Education, 60*(4), 15–23.
13. Huynh, T. P., Vo, D. H., & Hwang, C. L. (2018). Engineering and durability properties of eco-friendly mortar using cement-free SRF binder. *Construction and Building Materials, 160*, 145–155.
14. Jang, J. G., Ahn, Y. B., Souri, H., & Lee, H. K. (2015). A novel eco-friendly porous concrete fabricated with coal ash and geopolymeric binder: Heavy metal leaching characteristics and compressive strength. *Construction and Building Materials, 79*, 173–181.
15. Khan, M. M. R., & Islam, M. M. (2015). Materials and manufacturing environmental sustainability evaluation of apparel product: knitted T-shirt case study. *Textiles and Clothing Sustainability, 1*(1), 8.
16. Lopes, F. C., Tichota, D. M., Pereira, J. Q., Segalin, J., de Oliveira Rios, A., & Brandelli, A. (2013). Pigment production by filamentous fungi on agro-industrial byproducts: an eco-friendly alternative. *Applied Biochemistry and Biotechnology, 171*(3), 616–625.
17. Mahdavinejad, M., Ashtiani, S. R., Ebrahimi, M., & Shamshirband, M. (2013). Proposing a flexible approach to architectural design as a tool for achievement eco-friendly multi-purpose buildings. In *Advanced Materials Research* (Vol. 622, pp. 1856–1859). Trans Tech Publications.
18. Mateus, R., Neiva, S., Bragança, L., Mendonça, P., & Macieira, M. (2013). Sustainability assessment of an innovative lightweight building technology for partition walls–comparison with conventional technologies. *Building and Environment, 67*, 147–159.
19. Mirjalili, B. B. F., Bamoniri, A., & Akbari, A. (2011). Nano-BF 3· SiO 2: a reusable and eco-friendly catalyst for synthesis of quinoxalines. *Chemistry of Heterocyclic Compounds, 47*(4), 487.
20. Oglu Huseynov, E. F. (2011). Planning of sustainable cities in view of green architecture. *Procedia Engineering, 21*, 534–542.

21. Park, K. G., & Kim, J. (2018). Design Characteristics of Nature-Inspired Architecture. *Modern Environmental Science and Engineering, 4*(5), 381–414.

22. Ragheb, A., El-Shimy, H., & Ragheb, G. (2016). Green architecture: a concept of sustainability. *Procedia-Social and Behavioral Sciences, 216*, 778–787.

23. Raja, P. B., Qureshi, A. K., Rahim, A. A., Osman, H., & Awang, K. (2013). Neolamarckia cadamba alkaloids as eco-friendly corrosion inhibitors for mild steel in 1 M HCl media. *Corrosion Science, 69*, 292–301.

24. Rebah, F. B., & Siddeeg, S. M. (2017). Cactus an eco-friendly material for wastewater treatment: A review. *Journal of Materials and Environmental Sciences, 8*(5), 1770–1782.

25. Sekar, S. K. (2016). Mechanical and fracture characteristics of eco-friendly concrete produced using coconut shell, ground granulated blast furnace slag and manufactured sand. *Construction and Building Materials, 103*, 1–7.

26. Shabbir, M., Rather, L. J., Bukhari, M. N., Shahid, M., Khan, M. A., & Mohammad, F. (2016). An eco-friendly dyeing of woolen yarn by Terminalia chebula extract with evaluations of kinetic and adsorption characteristics. *Journal of Advanced Research, 7*(3), 473–482.

27. Singh, A., Sharma, P. K., & Malviya, R. (2011). Eco friendly pharmaceutical packaging material. *World Applied Sciences Journal, 14*(11), 1703–1716.

28. Stadel, A., Eboli, J., Ryberg, A., Mitchell, J., & Spatari, S. (2011). Intelligent sustainable design: Integration of carbon accounting and building information modeling. *Journal of Professional Issues in Engineering Education and Practice, 137*(2), 51–54.

29. Ukoba, O. K., Ogunkoya, A. K., & Soboyejo, W. (2011). Development of an eco-friendly Bamboo Bicycle. *Pacific Journal of Science and Technology, 12*(1), 102–108.

30. Vijayaraghavan, P., & Vincent, S. G. P. (2012). Cow dung as a novel, inexpensive substrate for the production of a halo-tolerant alkaline protease by Halomonas sp. PV1 for eco-friendly applications. *Biochemical Engineering Journal, 69*, 57–60.

31. Wang, X., Yu, R., Shui, Z., Song, Q., & Zhang, Z. (2017). Mix design and characteristics evaluation of an eco-friendly Ultra-High

Performance Concrete incorporating recycled coral based materials. *Journal of Cleaner Production*, *165*, 70–80.

32. Yun, R., Lu, Y., & Filip, P. (2010). Application of extension evaluation method in development of novel eco-friendly brake materials. *SAE International Journal of Materials and Manufacturing*, *2*(2), 1–7.

CHAPTER 8

Environmental Impact Assessment

CONTENTS

8.1. INTRODUCTION

Environmental Impact Assessment (EIA) can extensively be characterized as an investigation of the impacts of a proposed task, plan or program on the earth. The methodological, legitimate, and procedural establishments of EIA were built up in 1970 by the order of the National Environmental Policy Act (NEPA) in the US. At the worldwide level, loaning banks and two-sided help offices have EIA systems that apply to the recipient and borrowing nations (Rosenberg et al., 1986; Jay et al., 2007). Most developing regions have likewise grasped and are formalizing EIA through enactment. The paper features the advancement to current status, the legitimate structure, ideas, procedures and standards of EIA and related investigations (Stewart-Oaten et al., 1986; Lattemann and Höpner, 2008).

8.2. ORIGIN OF EIA

Prior to the First World War, fast urbanization and industrialization in western nations were causing quick loss of natural assets (Potting et al., 199; Morgan, 2012). This proceeded to the period after the Second World War offering ascends to worries for contamination, quality of life and environmental pressure. In the mid-60s, financial specialists and individuals understood that the activities they were undertaking were influencing the earth, resources, crude materials, and individuals. Subsequently, pressure teams were made with the point of getting an instrument that can be utilized to shield nature in any advancement (Branis and Kruzikova, 1994; Knox, 2002). The USA chose to react to these issues and set up a National Environmental Policy Act in 1970 to think about its objective as far as natural assurance was concerned. The USA turned into the principal nation to sanction enactment on EIA. This was the first occasion when that EIA turned into the official instrument to be utilized to secure nature. In 1972, in Stockholm, the UN Conference on the Environment and resulting spectacles dignified EIA. At present, every single developed nation has environmental laws while the greater part of the developing nations is as yet accepting it (Caldwell, 1988; Clarke, 2009). Multilateral and reciprocal loan specialists included EIA prerequisites in their undertaking qualification criteria (Pope et al., 2004).

8.3. EIA IN DEVELOPING NATIONS

Up to this point, EIA as another idea was not promptly comprehended and acknowledged as an instrument in developing nations. Engineers opposed and contended that it was hostile to advancement since laws and approaches supporting it directed that terrains improvements causing negative effects ought to be stopped (Lohani et al., 1997; Appiah-Opoku, 2001; Wood, 2003). More or less, EIA was viewed as simply one more bureaucratic hindrance in the way of advancement. Also, it was considered as a vile method by which industrialized countries mean to prevent developing nations from breaking the endless loop of destitution. Thirdly, the specialists in the developing nations were outsiders who were seen as operators of colonization. The requirement for EIAs has gotten progressively significant and is currently a statutory prerequisite in many creating nations (Wood, 2003; Alshuwaikhat, 2005).

Generally, the decision of new ventures was essentially founded on one rule: monetary practicality. Currently, a second and third decision criteria, natural and social effect, have become a solid measure, henceforth the triple main concern approach (financial, ecological and social) to extend suitability (Ebisemiju, 1993).

8.4. EIA IN UNIVERSAL ENVIRONMENTAL LAW PERSPECTIVE

Main Multilateral Environmental Agreements (MEAs) have perceived review and developments in EIA legitimate, procedure and official arrangements. The vital agreements have been discoursed below.

8.4.1. Agreement on Environmental Impact Assessment in a Multi-National Perspective (Espoo, 1991)

This is the foremost multi-parallel EIA bargain. It takes a gander at EIA in a trans-limit setting and went into power in 1997. The Espoo Convention forms lay down commitments of Parties to survey the natural effect of specific exercises at a beginning period of arranging (Koyano, 2008; Zharova, 2012). It likewise sets out the general commitment of states to advise and counsel each other on every significant task viable that is probably going to have a noteworthy unfavorable ecological effect across the outskirts (Marsden, 2010; Stokke and Thommessen, 2013).

Aside from stipulating the obligation of signatory nations concerning recommendations that have trans boundary impacts, it depicts the standards, arrangements, techniques to be followed and a rundown of exercises, substance of documentation and standards of immensity that apply (Robinson, 1991; Schrage and Bonvoisin, 2008; Đereg, 2011).

8.4.2. Rio Declaration (1992)

Rule 17 of the Rio Declaration on Environment and Development (RDED) calls for utilization of EIA as a nationwide simple leadership instrument to be utilized in surveying whether proposed exercises are probably going to have a noteworthy antagonistic effect on the earth (Batt and Short, 1992; Weiss, 1992). It likewise underscored the job of an able national expert in the basic leadership process. The other standard of this presentation that is important to EIA drill is the use of the prudent guideline (Porras, 1992; Handl, G. (2012).

Agenda 21, which was likewise because of this agreement, recommends that administrations should:

i. *"Promote the development of appropriate methodologies for making integrated energy, environment and economic policy decisions for sustainable development, inter alia, through environmental impact assessment (9.12(b))*

ii. *Develop, improve and apply environmental impacts assessment, to foster sustainable industrial development (9.18)*

iii. *Carry out investment analysis and feasibility studies including environmental assessments for establishing forest based processing enterprises.*

iv. *Introduce appropriate EIA procedures for proposed projects likely to have significant impacts upon biological diversity, providing for suitable information to be made widely available and for public participation, where appropriate, and encourage the assessment of impacts of relevant policies and programs on biological diversity (15.5(k)"* (UNICED 1992).

Agenda 21 defines the outline within which nations can institute their nationwide ecological regulations.

a. **UN Convention on environmental change and Biological Diversity (1992)** referred to EIA as an actualizing instrument of these shows (articles 4 and 14 individually).

b. **Doha Ministerial Declaration** urges nations to impart ability and experience to individuals wishing to implement ecological surveys at the domestic level (November 2001).

c. **UNECE (Aarhus) Convention on Access to Information, Public Participation in Decision Making and Access to Justice in Environmental Matters (1998)** refuges the choices at the degree of activities and plans, projects and arrangements and by augmentation, applies to EIA and SEA.

d. **UN Conference** on the Environment in Stockholm 1972.

8.4.3. Multidimensional and Mutual Economic Institutions Ecological Safeguards

Venture banks like Asian Development Bank (ADB), African Development Bank (AfDB), European Investment Bank (EIB), European Bank for Reconstruction and Development (EBRD), World Bank (WB), Japanese Bank for International Cooperation (JBIC) have ecological shields to guarantee that financing of activities is not just founded on the prudent standard, deterrent activity as opposed to corrective treatment yet supportable improvement (WBCSD, 2005). In spite of the fact that their operational arrangements and prerequisites differ in specific regards, the advancement banks adhere to a generally standard strategy for the planning and endorsement of an EIA report (Gonzalo et al., 2017; Pascual et al., 2018). Acquiring nations are answerable for the arrangement of the EIA, and this necessity potentially more than some other has impacted the presentation and advancement of EIA in many developing nations. The EIA ought to inspect venture choices and recognize methods for improving undertaking determination, siting, arranging, structure and usage by forestalling, limiting, moderating and making up for unfavorable ecological effects (Rao, 2010). Much the same as different banks, the World Bank has standards for screening ventures as given below (Morgan, 2012).

8.4.3.1. Category A

In the event that the undertaking prone to have noteworthy ecological effects that are delicate, differing or phenomenal. These effects may influence a territory more extensive than the networks profiting by framework ventures.

8.4.3.2. Category B

In the event that the activities potential antagonistic ecological effects on human populaces or earth significant zones are less unfriendly than those of Category A tasks. These effects are site-explicit; scarcely any of them are permanent; and as a rule alleviation measures can be structured more promptly than for Category A ventures.

8.4.3.3. Category C

On the off chance that the undertaking is probably going to have insignificant or no unfavorable ecological effects. When the undertaking is surveyed and decided as Category C, no further activity would be essential. A few instances of Category C ventures include Education (for example limit-building, and so forth, excluding school development), Family planning and so on.

All tasks financed by the Banks ought to likewise consent to the necessities of pertinent multilateral ecological understandings (MEA) to which the host nation is a gathering, including the Montreal Protocol (on ozone draining substances), the UN Convention on Climate Change and the Kyoto Protocol (on ozone-depleting substance emanations) and the Aarhus Convention (on natural data). Every single worldwide association and respective offices as often as possible update their systems and it is essential to get the present rendition from the association (Palerm, 1999).

8.4.4. National Legislations

National enactment may incorporate a statutory necessity for an EIA to be done in a recommended way for explicit improvement exercises. Most enactment records ventures for which EIA is a compulsory prerequisite. The statutory necessity to do an EIA for explicit undertakings will, for instance, require enrolled specialists to do the investigation, the authority with the assistance of lead offices and specialized panels to survey the EIA and endorse the task (Pardo, 1997).

Other national lawful necessities that oversee the utilization and insurance of assets like water, fisheries, timberlands, natural life, general wellbeing and so forth should be distinguished and agreed to during an EIA.

8.4.5. Institutional Framework

EIA institutional frameworks change from nation-to-nation and reflecting various sorts of administration. In certain nations, either the Ministry of Environment or an assigned position or Planning Agency control EIA (Stærdahl et al., 2004; Esteves et al., 2012).

Ecological issues likewise include numerous orders and numerous administration bodies with general natural and asset management laws. Information will in this manner be gathered and grouped from a wide scope of specialized services, other government specialists and parastatals where pertinent (Borzel, 2000; Victor and Agamuthu, 2014).

8.5.TERMS OF REFERENCE (TOR) PREPARATION

ToR lays down what is anticipated from a professional or an expert when doing an EIA. ToRs can be basic or expound yet expand ToRs are generally not prescribed. There are no general organizations for terms of reference, which will be reasonable for each investigation. In any case, there are general standards, which ought to be seen while planning ToR for the EIA (Tomlinson and Atkinson, 1987; Ebisemiju, 1993).

i. The ToR ought, to begin with, a short depiction of the program or undertaking. This ought to incorporate an arrangement of the territory that will be influenced either by implication or legitimately (Brismar, 2004).

ii. The study ought to guarantee that the advisors or experts center around the significant issues and the most genuine likely effects recognized during checking for example air outflow, squander water release and so forth. The open doors for improving any positive advantages from the undertaking ought to likewise be featured. This part of ToR is generally submitted to the assigned expert for investigation and endorsement (Nadeem and Hameed, 2008).

iii. The ToR ought to contain rapid references to which defend arrangements might be important and which lawful prerequisites ought to be applied.

iv. The ToR should give a sign of the group considered important for the examination and a group chief distinguished. Contingent upon the extent of the investigation this might be multi-disciplinary. Notwithstanding, as the group, ought not to be formally forced on the specialist.

v. If global specialists are doing the EIA, it is critical to make arrangements for neighborhood limits working in the ToR. Aside from empowering in-nation aptitude to be developed, this will advance greater association and comprehension of the issues raised by the investigation. As most EIA contemplates are of a moderately brief span, this is presumably best accomplished through the connection of venture defender to the experts during the examination or an emphasis on the utilization of neighborhood staff workforce for a portion of the assignments (Spaling and Smit, 1993).

vi. The expected date of initiation and time breaking point ought to be given and expert program of work must be inside the given time limit.

vii. The spending farthest point ought to be given in the ToR. The sort of specialists, and whether outside or nearby, and the term of their sources of info will, for the most part, be the choosing cost factors albeit a huge field study or estimation program with research facility investigation could altogether expand costs. Any help to be given by the Client to decrease expenses ought to be openly expressed in the ToR.

viii. Consultant installments proposition ought to be made and attached to explicit achievements, for example, the expert will be paid 20% of their payment of draft report by the customer and so on.

ix. Reporting necessities ought to be unmistakably expressed and ought to agree to nearby or global announcing rules. The arrangement of EIS must be clear and the number of duplicates in digital and hard should be expressed.

x. ToR should make arrangements for the specialists to develop the terms of reference so as to improve the nature of EIA.

8.6. ENVIRONMENTAL IMPACT ASSESSMENT (EIA) PROCESS

The primary part of an ecological evaluation is called an Initial Environmental Examination (IEE) and the subsequent is Environmental Impact Studies (EIS) or thorough EIA (Saarikoski, 2000; Ramanathan, 2001).

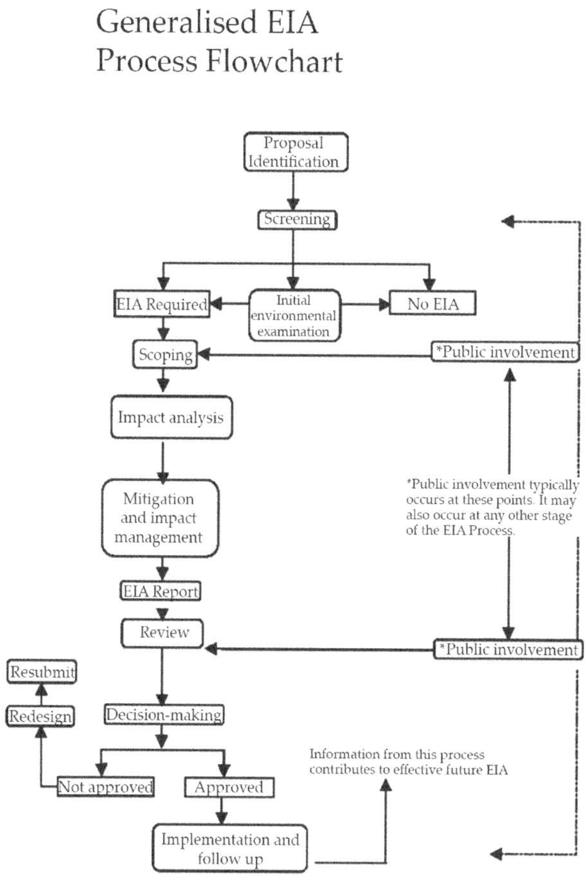

Figure 8.1: Comprehensive EIA route flowchart.

[Source: https://orkustofnun.is/gogn/unu-gtp-sc/UNU-GTP-SC-10–0801.pdf]

IEE is completed to decide if possibly antagonistic ecological impacts are huge or whether moderation measures can be embraced to decrease or wipe out these unfavorable impacts. The IEE contains a short proclamation of key ecological issues, in view of promptly accessible data, and is utilized in the early (pre-plausibility) period of venture arranging. The IEE likewise recommends whether top to bottom investigations are required. At the point when an IEE can give a positive answer to natural issues, an EIA is not vital. IEE additionally requires master guidance and specialized contribution from natural experts with the goal that potential ecological issues can be obviously characterized (Jay et al., 2007; Morgan, 2012).

EIA is a technique used to look at the ecological outcomes or effects, both gainful and unfriendly, of a proposed advancement venture and to guarantee that these impacts are considered in a venture plan (Sallenave, 1994). The EIA is in this manner dependent on forecasts. These effects can incorporate every single applicable part of the natural, financial, social, and human condition. The examination consequently requires a multidisciplinary approach and ought to be done right on time at the plausibility phase of an undertaking. At the end of the day, an undertaking ought to be evaluated for its ecological practicality (Tukker, 2000; Steinemann, 2001).

EIA ought to in this way be seen as an indispensable piece of the task arranging process. In contrast to the ecological review (EA), which is led to existing undertakings, the EIA is applied to new ventures and the development parts of existing tasks. The sections of an EIA from screening to catch up are delineated in Figure 8.1.

8.6.1. Screening

EIA process commences with venture screening. Screening is completed to decide if a proposition ought to be liable to EIA and, assuming this is the case, at what level of thoroughness. Rules for whether an EIA is required are nation explicit relying upon the laws or standards of activity. Enactment frequently determines the standards for screening and complete EIA. Improvement banks likewise screen ventures introduced for financing to choose whether an EIA is required utilizing their set criteria (Weston, 2000; Koornneef et al., 2008).

The yield of the screening procedure is commonly an archive called an Initial Environmental Examination or Evaluation (IEE). The principle decision will be an arrangement of the task as indicated by its probable natural affectability. This will decide if an EIA is required and provided that this is true, to what extent (Arts et al., 2001).

8.6.2. Scoping

The point of EIA is not to do thorough investigations on every ecological effect for all ventures. Checking is utilized to distinguish the key issues of worry at the beginning period in the organizing procedure (Ahmed and Sammy, 1987). The aftereffects of this process will decide the degree, profundity, and terms of reference to be tended to inside the Environmental articulation (Slootweg and Kolhoff, 2003; Nadeem and Hameed, 2008). Scoping is performed to:

i. Classify issues and concerns for thought in an EIA.
ii. Ensure a pertinent EIA.
iii. Enable those liable for an EIA investigation to appropriately concise the investigation group on the other options and on effects to be considered at various degrees of examination.
iv. Determine the appraisal strategies to be utilized.
v. Identify every influenced intrigue.
vi. Provide an open door for public association in deciding the elements to be surveyed, and encourage early concession to argumentative issues.
vii. Save money and time.
viii. Institute terms of reference (TOR) for EIA investigation.

Scoping ought to be a progressing exercise over the span of the task. The accompanying natural apparatuses can be utilized in the scoping exercise.

8.6.2.1. Checklists

Checklists are typical arrangements of the sorts of effects related to a specific kind of undertaking. Checklists' strategies are essential for sorting out data or guaranteeing that no potential effect is ignored. They include list inquiries to highlight the task and environmental impacts. They are conventional in nature and are utilized as bits of help in the appraisal.

8.6.2.2. Matrices

Matrix strategies recognize associations between different task activities and ecological parameters and segments. They involve a rundown of task exercises with an agenda of ecological parts that may be influenced by these exercises. A network of potential cooperation is delivered by joining these two records (putting one on the vertical hub and the other on the flat hub). They ought to ideally cover both the development and the activity periods of the undertaking, on the grounds that occasionally, the previous causes more prominent effects than the last mentioned. Be that as it may, matrices likewise have their impediments: they don't unequivocally speak to spatial or fleeting contemplations, and they do not enough discourse synergistic effects.

8.6.2.3. Networks

These are cause-impact stream outlines used to help in following the web connections that exist between various exercises related to activity and the natural framework with which they communicate. They are likewise significant in distinguishing immediate and combined effects. They are increasingly unpredictable and need skills for their powerful use.

8.6.2.4. Consultations

With leaders, influenced networks, ecological intrigue gatherings to guarantee that every single potential effect is recognized. Anyway, there can be peril in this when over the top discussion is done and some baseless effects remembered for the ToR.

8.6.3. Baseline Data Gathering

The expression "baseline "alludes to the assortment of foundation data on the biophysical, social and monetary settings proposed venture region. Ordinarily, data is gotten from possible sources, or the securing of new data through field samplings, meetings, overviews, and counsels with the general population. The assignment of gathering baseline data begins directly from the time of venture beginning; be that as it may, a larger part of this errand might be embraced during scoping and real EIA.

Baseline data is gathered for two primary purposes

i. To give a portrayal of the present status and patterns of natural elements (e.g., air contamination centralizations) of the host-zone against which anticipated changes can be thought about and assessed as far as noteworthiness, and

ii. To give a method for distinguishing genuine change by observing once a venture has been started

Just baseline data expected to help forecast the effects enclosed in the ToR and scoping reports ought to be gathered.

8.6.4. Impact Analysis and Extrapolation

Anticipating the size of an advancement likely effects and assessing their criticalness is the center of the ecological evaluation process (Morris and Therivel, 1995). Forecast ought to be founded on the accessible ecological baseline of the task territory. Such forecasts are depicted in quantitative or subjective terms.

8.6.5. Considerations in Impact Extrapolation

8.6.5.1. Scale of Impact

This is characterized by the seriousness of every potential effect and demonstrates whether the effect is permanent or rescindable and evaluated the potential pace of recuperation. The scale of an effect cannot be viewed as high if a significant antagonistic effect can be relieved.

8.6.5.2. Range of Impact

The spatial degree or the zone of impact of the effect ought to consistently be resolved. An effect can be site-explicit or constrained to the undertaking zone; a locally happening effect inside the territory of the proposed venture; a provincial effect that may stretch out past the neighborhood a national effect influencing assets on a national scale and once in a while trans-limit impacts, which may be worldwide.

8.6.5.3. Period of Impact

Ecological effects have a worldly measurement and should be reflected in an EIA. The effects emerging at various periods of the undertaking sequence may need to be measured. An effect that by and large goes on for just three to nine years after undertaking culmination might be named present moment. An effect, which proceeds for 10 to 20 years, might be characterized as medium-term, and effects that last past 20 years are considered as long haul.

8.6.5.4. Importance of the Impact

This alludes to the worth or measure of the effect. When an effect has been anticipated, its noteworthiness must be assessed utilizing a proper selection of criteria. The most significant types of basis are:

i. Specific legitimate prerequisites, for example, national regulations, models, worldwide understandings and conventions, significant approaches and so on.
ii. Public perspectives and objections.
iii. The threat to subtle biological systems and assets, for example,

can prompt elimination of species and consumption of assets, which can result, in clashes.

iv. The geographical range of the effect, for example, has trans-limit inferences.

v. Cost of alleviation.

vi. Duration (timeframe over which they will happen).

vii. Probability or likelihood of event (likely, improbable, and so on).

viii. Reversibility of effect (characteristic recuperation or helped by human mediation).

ix. Umber (and attributes) of individuals prone to be influenced and their areas.

x. Cumulative effects, for example, adding more effects to prevailing ones.

xi. Uncertainty in expectation because of the absence of precise information or complex frameworks. The prudent standard is upheld in this situation.

8.6.6. Impact extrapolation methodologies

A few systems can be utilized in anticipating the effects. The decisions ought to be fitting to the conditions. These can be founded on:

i. Professional decisions with sufficient thinking and supporting information. This strategy requires a high expert experience;

ii. Tests or Experiments (these can be costly);

iii. Previous experience;

iv. Numerical estimations and mathematical simulations. These can require a great deal of information and competency in numerical demonstrating without which shrouded mistakes can emerge Physical or optical examination. Point by point portrayal is expected to show the effect;

v. Geographical data frameworks;

vi. Risk calculation; and

vii. Economic assessment of environmental effects.

8.6.7. Investigation of Alternatives

Investigation of alternative is done to build up the liked or most ecologically solid, monetarily possible and gentle alternative for accomplishing venture targets.

The World Bank orders require the deliberate correlation of proposed speculation configuration as far as the site, innovation, forms and so forth as far as their effects and attainability of their alleviation, capital, repetitive costs, appropriateness under neighborhood conditions and institutional, preparing and observing prerequisites (World bank 1999). For every other option, the ecological expense ought to be measured to the degree conceivable and monetary qualities joined where attainable, and the essential for chose elective expressed. The investigation of options ought to incorporate NO PROJECT elective.

8.6.8. Moderation and Impact Organization

Moderation is done to keep away from, limit or counterbalance anticipated antagonistic effects and, where proper, to fuse these into an ecological administration plan or framework. For every potential unfriendly effect, the arrangement for its alleviation at each phase of the task ought to be recorded and cost, as this is significant in the determination of the favored other option.

The goals of moderation in this way are to:

i. Find better other options and methods for getting things done;
ii. Enhance the ecological and social advantages of an undertaking;
iii. Avoid, limit or cure unfavorable effects; and
iv. Ensure that lingering antagonistic effects are kept inside adequate levels.

Table 8.1: Design of Moderation Measures

Approach	Examples
Avoid	Channels, precipitators, clamor obstructions, dust, walled in areas, visual screening, untamed life passages, and changed time of exercises.
Replace	Recover comparative territory of identical environmental incentives in various areas.
Reduce	Filters, precipitators, noise barriers, dust, enclosures, visual screening, wildlife corridors, and changed time of activities.

Restore	Site rebuilding after development.
Compensate	Migration of dislodged networks, offices for the influenced networks, money related compensation for the influenced people and so on.

REFERENCES

1. Alshuwaikhat, H. M. (2005). Strategic environmental assessment can help solve environmental impact assessment failures in developing countries. *Environmental Impact Assessment Review*, *25*(4), 307–317.
2. Appiah-Opoku, S. (2001). Environmental impact assessment in developing countries: the case of Ghana. *Environmental Impact Assessment Review*, *21*(1), 59–71.
3. Arts, J., Caldwell, P., & Morrison-Saunders, A. (2001). Environmental impact assessment follow-up: good practice and future directions—findings from a workshop at the IAIA 2000 conference. *Impact Assessment and Project Appraisal*, *19*(3), 175–185.
4. Batt, J., & Short, D. C. (1992). The Jurisprudence of the 1992 Rio Declaration on Environment and Development: A Law, Science, and Policy Explication of Certain Aspects of the United Nations Conference on Environment and Development. *J. Nat. Resources & Envtl. L.*, *8*, 229.
5. Borzel, T. A. (2000). Why there is no'southern problem.' On environmental leaders and laggards in the European Union. *Journal of European Public Policy*, *7*(1), 141–162.
6. Branis, M., & Kruzikova, E. (1994). The environmental impact assessment act in the Czech Republic: Origins, introduction, and implementation issues. *Environmental Impact Assessment Review*, *14*(2–3), 195–201.
7. Brismar, A. (2004). Attention to impact pathways in EISs of large dam projects. *Environmental Impact Assessment Review*, *24*(1), 59–87.
8. Caldwell, L. K. (1988). Environmental impact analysis (EIA): origins, evolution, and future directions. *Impact Assessment*, *6*(3–4), 75–83.
9. Clarke, R. (2009). Privacy impact assessment: Its origins and development. *Computer Law & Security Review*, *25*(2), 123–135.
10. Đereg, N. (2011). Environmental impact assessment in a trans-boundary context in the SEE countries. In *Environmental Security in South-Eastern Europe* (Vol. 1, pp. 183–196). Springer, Dordrecht.
11. Ebisemiju, F. S. (1993). Environmental impact assessment: making it work in developing countries. *Journal of Environmental Management*, *38*(4), 247–273.
12. Esteves, A. M., Franks, D., & Vanclay, F. (2012). Social impact assessment: the state of the art. *Impact Assessment and Project*

Appraisal, 30(1), 34–42.

13. Gonzalo, J., Zewdie, S., Tenkir, E., & Moges, Y. (2017). REDD+ and carbon markets: the Ethiopian process. In *Managing Forest Ecosystems: The Challenge of Climate Change* (Vol. 1, pp. 151–183). Springer, Cham.
14. Handl, G. (2012). Declaration of the United Nations conference on the human environment (Stockholm Declaration), 1972 and the Rio Declaration on Environment and Development, 1992. *United Nations Audiovisual Library of International Law, 11, 1–13.*
15. Jay, S., Jones, C., Slinn, P., & Wood, C. (2007). Environmental impact assessment: Retrospect and prospect. *Environmental Impact Assessment Review, 27*(4), 287–300.
16. Knox, J. H. (2002). The myth and reality of transboundary environmental impact assessment. *American Journal of International law, 96*(2), 291–319.
17. Koornneef, J., Faaij, A., & Turkenburg, W. (2008). The screening and scoping of environmental impact assessment and strategic environmental assessment of carbon capture and storage in the Netherlands. *Environmental Impact Assessment Review, 28*(6), 392–414.
18. Koyano, M. (2008). The significance of the Convention on Environmental Impact Assessment in a Transboundary Context (Espoo Convention) in international environmental law: examining the implications of the Danube Delta case. *Impact Assessment and Project Appraisal, 26*(4), 299–314.
19. Lattemann, S., & Höpner, T. (2008). Environmental impact and impact assessment of seawater desalination. *Desalination, 220*(1–3), 1–15.
20. Lohani, B. N., Evans, J. W., Everitt, R. R., Ludwig, H., Carpenter, R. A., & Tu, S. L. (1997). *Environmental Impact Assessment For Developing Countries in Asia: Volume 1-Overview* (Vol. 1, pp. 1–30). Asian Development Bank.
21. Marsden, S. (2010). Developing Approaches to Trans-boundary Environmental Impact Assessment in China: Co-operation through the Greater Tumen Initiative and in the Pearl River Delta Region. *Chinese Journal of International Law, 9*(2), 393–414.
22. Morgan, R. K. (2012). Environmental impact assessment: the state of the art. *Impact Assessment and Project Appraisal, 30*(1), 5–14.

23. Nadeem, O., & Hameed, R. (2008). Evaluation of environmental impact assessment system in Pakistan. *Environmental Impact Assessment Review*, *28*(8), 562–571.
24. Palerm, J. R. (1999). Public participation in environmental impact assessment in Spain: three case studies evaluating national, Catalan and Balearic legislation. *Impact Assessment and Project Appraisal*, *17*(4), 259–271.
25. Pardo, M. (1997). Environmental impact assessment: Myth or reality? Lessons from Spain. *Environmental Impact Assessment Review*, *17*(2), 123–142.
26. Pascual, U., Garmendia, E., Phelps, J., & Ojea, E. (2018). Opportunities and Conditions for Successful Foreign Aid to the Forestry Sector. In *Aid Effectiveness for Environmental Sustainability* (Vol. 1, pp. 257–305). Palgrave Macmillan, Singapore.
27. Pope, J., Annandale, D., & Morrison-Saunders, A. (2004). Conceptualising sustainability assessment. *Environmental Impact Assessment Review*, *24*(6), 595–616.
28. Porras, I. M. (1992). The Rio declaration: a new basis for international co-operation. *Rev. Eur. Comp. & Int'l Envtl. L.*, *1*, 245.
29. Potting, J., Schöpp, W., Blok, K., & Hauschild, M. (1998). Comparison of the acidifying impact from emissions with different regional origin in life-cycle assessment. *Journal of Hazardous Materials*, *61*(1–3), 155–162.
30. Ramanathan, R. (2001). A note on the use of the analytic hierarchy process for environmental impact assessment. *Journal of Environmental Management*, *63*(1), 27–35.
31. Rao, P. K. (2010). Green Economic Policies: Regional and Global Levels. In *The Architecture of Green Economic Policies* (Vol. 1, pp. 123–143). Springer, Berlin, Heidelberg.
32. Robinson, N. A. (1991). International trends in environmental impact assessment. *BC Envtl. Aff. L. Rev.*, *19*, 591.
33. Rosenberg, D. M., Danks, H. V., & Lehmkuhl, D. M. (1986). Importance of insects in environmental impact assessment. *Environmental Management*, *10*(6), 773–783.
34. Saarikoski, H. (2000). Environmental impact assessment (EIA) as collaborative learning process. *Environmental Impact Assessment Review*, *20*(6), 681–700.

35. Sallenave, J. (1994). Giving traditional ecological knowledge its rightful place in environmental impact assessment. *Northern Perspectives*, *22*(1), 16–19.
36. Schrage, W., & Bonvoisin, N. (2008). Transboundary impact assessment: frameworks, experiences and challenges. *Impact Assessment and Project Appraisal*, *26*(4), 234–238.
37. Slootweg, R., & Kolhoff, A. (2003). A generic approach to integrate biodiversity considerations in screening and scoping for EIA. *Environmental Impact Assessment Review*, *23*(6), 657–681.
38. Spaling, H., & Smit, B. (1993). Cumulative environmental change: conceptual frameworks, evaluation approaches, and institutional perspectives. *Environmental Management*, *17*(5), 587–600.
39. Stærdahl, J., Schroll, H., Zakaria, Z., Abdullah, M., Dewar, N., & Panich, N. (2004). Environmental impact assessment in Malaysia, South Africa, Thailand, and Denmark: Background, layout, context, public participation and environmental scope. *The Journal of Transdisciplinary Environmental Studies*, *3*(1), 1–19.
40. Steinemann, A. (2001). Improving alternatives for environmental impact assessment. *Environmental Impact Assessment Review*, *21*(1), 3–21.
41. Stewart-Oaten, A., Murdoch, W. W., & Parker, K. R. (1986). Environmental impact assessment: "Pseudoreplication" in time?. *Ecology*, *67*(4), 929–940.
42. Stokke, O. S., & Thommessen, O. B. (2013). Convention on Environmental Impact Assessment in a Transboundary Context, Espoo, 1991. In *Yearbook of International Cooperation on Environment and Development 2003–04* (Vol. 1, pp. 84–85). Routledge.
43. Tomlinson, P., & Atkinson, S. F. (1987). Environmental audits: proposed terminology. *Environmental Monitoring and Assessment*, *8*(3), 187–198.
44. Tukker, A. (2000). Life cycle assessment as a tool in environmental impact assessment. *Environmental Impact Assessment Review*, *20*(4), 435–456.
45. Victor, D., & Agamuthu, P. (2014). Policy trends of strategic environmental assessment in Asia. *Environmental Science & Policy*, *41*, 63–76.

46. Weiss, E. B. (1992). Environment and trade as partners in sustainable development: A commentary. *American Journal of International Law, 86*(4), 728–735.

47. Weston, J. (2000). EIA, decision-making theory and screening and scoping in UK practice. *Journal of Environmental Planning and Management, 43*(2), 185–203.

48. Wood, C. (2003). Environmental impact assessment in developing countries. *International Development Planning Review, 25*(3), 301–321.

49. Wood, C. (2003, November). Environmental impact assessment in developing countries: an overview. In *Conference on New Directions in Impact Assessment for Development: Methods and Practice* (Vol. 24, pp. 11–41). Manchester, United Kingdom: EIA Centre School of Planning and Landscape University of Manchester.

50. Zharova, L. (2012). Integrated management as a background of modern environmental policy. *Global Business and Management Research: An International Journal, 4*(2), 123–131.

INDEX

T

U

V

W

Printed in the United States
By Bookmasters